Modeling and Analysis of Eclipsing Binary Stars

The theory and design principles of PHOEBE

AAS | IOP Astronomy

AAS Editor in Chief

Ethan Vishniac, John Hopkins University, Maryland, US

About the program:

AAS-IOP Astronomy ebooks is the official book program of the American Astronomical Society (AAS), and aims to share in depth the most fascinating areas of astronomy, astrophysics, solar physics and planetary science. The program includes publications in the following topics:

- GALAXIES AND COSMOLOGY
- INTERSTELLAR MATTER AND THE LOCAL UNIVERSE
- STARS AND STELLAR PHYSICS
- EDUCATION, OUTREACH AND HERITAGE
- HIGH-ENERGY PHENOMENA AND FUNDAMENTAL PHYSICS
- THE SUN AND THE HELIOSPHERE
- THE SOLAR SYSTEM, EXOPLANETS, AND ASTROBIOLOGY
- INSTRUMENTATION, SOFTWARE, LABORATORY ASTROPHYSICS AND DATA

Books in the program range in level from short introductory texts on fast-moving areas, graduate and upper-level undergraduate textbooks, research monographs and practical handbooks.

For a complete list of published and forthcoming titles, please visit iopscience.org/books/aas.

About the American Astronomical Society

The American Astronomical Society (aas.org), established 1899, is the major organization of professional astronomers in North America. The membership (~7,000) also includes physicists, mathematicians, geologists, engineers and others whose research interests lie within the broad spectrum of subjects now comprising the contemporary astronomical sciences. The mission of the Society is to enhance and share humanity's scientific understanding of the universe.

Modeling and Analysis of Eclipsing Binary Stars

The theory and design principles of PHOEBE

Andrej Prša
Department of Astrophysics and Planetary Science, Villanova University, Mendel Science Center, Villanova PA, USA

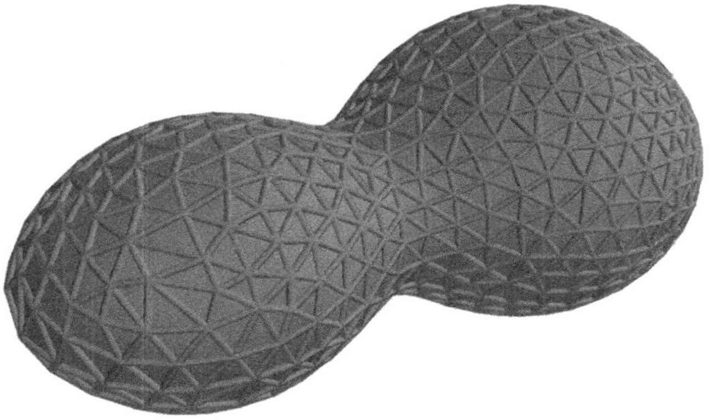

The Theeory and Design Priniciples of PHOEBE

IOP Publishing, Bristol, UK

Supplementary material is available at http://iopscience.iop.org/book/978-0-7503-1287-5.

Multimedia content is available at http://iopscience.iop.org/book/978-0-7503-1287-5.

ISBN 978-0-7503-1287-5 (ebook)
ISBN 978-0-7503-1288-2 (print)
ISBN 978-0-7503-1289-9 (mobi)

DOI 10.1088/978-0-7503-1287-5

Version: 20181201

AAS–IOP Astronomy
ISSN 2514-3433 (online)
ISSN 2515-141X (print)

British Library Cataloguing-in-Publication Data: A catalogue record for this book is available from the British Library.

Published by IOP Publishing, wholly owned by The Institute of Physics, London

IOP Publishing, Temple Circus, Temple Way, Bristol, BS1 6HG, UK

US Office: IOP Publishing, Inc., 190 North Independence Mall West, Suite 601, Philadelphia, PA 19106, USA

To Kelly, Dominika, Anja, and Celestina, my shining stars that light my way through life.

Dedicated to the loving memory of Maks Simončič†, 1931–2017

Contents

Preface

Welcome to the exciting world of eclipsing binary stars! Of course, when I say "eclipsing binary stars," what I really mean is the world of fundamental stellar parameters, cosmic distance scales, calibrations of evolutionary and stellar population theories, the world of extrasolar planets, pulsating stars, degenerate remnants, dwarfs and giants, the world of ambitious observing surveys, extensive supercomputer calculations, and big data science. Eclipsing binaries span all these fields of study thanks to their unique advantage that makes them indispensable astrophysical laboratories: they enable us to study the structure and properties of their stellar (and substellar) components unencumbered by external calibrations. At the same time, eclipsing binaries do not discriminate among the types of components they host—we can find anything from black holes to supergiants. Transiting extrasolar planets are eclipsing binaries too; they just happen to have an extreme mass ratio. To top that off, consider that the majority of all stars (save the lowest M types) are found in binary and multiple systems; studying eclipsing binaries in their own right means that we study *commonplace* systems rather than oddball single stars.

Let me take a step back and state this in a slightly different way. A thorough understanding of fundamental stellar parameters (masses, radii, luminosities, ages, chemical compositions, and distances) and processes (energy transport mechanisms, nucleosynthesis) in stars across the Hertzsprung–Russell diagram is the core of many fields in astrophysics. To rigorously study stars, we need to determine their properties as accurately as possible, using the minimum number of underlying assumptions. Eclipsing binary stars are ideal objects to achieve this goal: the favorable alignment of the line of sight with the orbital plane and the basic principles of classical dynamics that govern the motion of the components in a binary reduce the determination of principal parameters to a tractable geometric problem. This unique property, in contrast with other means of determining stellar radii that either apply to only a handful of objects (such as resolving the disk of a star) or are plagued with a larger uncertainty (i.e., period–luminosity relationships), makes eclipsing binaries the key calibrators of stellar properties and distance gauges.

In this book I will talk only about eclipsing binary stars. I will defer to you, the reader, how you will use what eclipsing binaries give you in your own field of expertise. I will not assume that I know the intricacies and best practices of your own field. Throughout the book I will assume that you are interested in what eclipsing binaries can do for you, which means that we are taking this trip of exploration together. Thus, except in this opening part of the book, I will use first-person plural ("we").

This is not the first book on eclipsing binary stars. At least three recent books come to mind that specialize in this topic: Josef Kallrath and Gene Milone's *Eclipsing Binary Stars: Modeling and Analysis*, currently in its second edition; Peter Eggleton's *Evolutionary Processes in Binary and Multiple Stars*; and Ronald Hilditch's *An Introduction to Close Binary Stars*. Of course, I cannot omit the classics, most notably Robert Aitken's *The Binary Stars* from 1935 and Zdeněk Kopal's *Close Binary Systems* from 1959. In addition, there are numerous conference proceedings, too many to list here, but thankfully we live in an information era where the complete list is a few clicks away.

In this book I focus on the topics that I found challenging to master and for which I thought the information was scattered over many other books and papers. I do my best to provide relevant references as I go along, so in that respect this is more of a very long review paper than a book. I will interchangeably use the more formal style of writing and a more relaxed one, depending on the dryness of the discussed topic.

To truly understand the ins and outs of eclipsing binary stars, I would need to master swaths of classical physics, modern physics, quantum mechanics, statistical thermodynamics, observational astrophysics, astrostatistics, and big data science, as well as sociology and psychology (in order to inform my family that I need yet another late night to work on the book). No one can live up to that, and I will not even attempt to do that. I thus immediately disclose that the book is *bound* to be inaccurate and sloppy at times—mostly when I learn just enough to give myself the impression that I actually understand the underlying mechanisms. I have been in the field for almost 20 years, but there is still so much to learn, to explore, to improve. Because of that, here is what I consider the most exciting aspect of this project: take this book as a living document. Even as the book is published, I will continue to work on it as I learn more, and I kindly invite you to help me with it: let me know what parts of the book are unclear, what parts are plain wrong, and what topics are missing altogether.

This work describes the logic and principles that went into building the eclipsing binary modeling code PHOEBE. PHOEBE stands for PHysics Of Eclipsing BinariEs. Yes, it is a bit of a stretch. The name goes back to the time when my brief thought of bestowing that name onto my sometime-in-the-future daughter met with an instant veto. This, however, is the only mention of PHOEBE, as I did my best to present the principles for their own worth rather than attach them to a certain modeling code.

In the book I inevitably self-plagiarize a bit. While I did my best to rewrite and expand on the material presented in the papers I authored and coauthored, there have been instances where I spent a lot of time in the past trying to figure out the best way to convey a point, and I felt that reinventing the way to say the same thing would have been counterproductive. At other times I was working on the book and the paper contemporaneously and it seemed natural to include the write-up in both places. I did my best not to plagiarize anyone else's work, though, and any similarity with other works is unintentional.

To conclude, allow me to thank you, the reader, for the time and effort you put into reading this book. To appreciate the beauty of the field of eclipsing binary stars is no walk in the park: the learning slope is steep. I tried to keep the length of the book unintimidating while providing content that is informative and complete. I would thoroughly appreciate a word on whether I have succeeded. I certainly hope that you will emerge on the other end richer for the experience and appreciative of the intricacies of the eclipsing binary field.

Andrej Prša
Villanova, March 2018

Acknowledgments

My first and foremost thanks go to my lovely wife Kelly—for being my scientific pillar, my familial support, my strength, and my inspiration: I could not have done this without you!

To our daughters Dominika, Anja, and (about to be born) Celestina—for making this world a better place and for making our lives complete!

To my friends and colleagues Kyle Conroy, Martin Horvat, Angela Kochoska, Dave Jones, Bert Pablo, Joe Giammarco, Steven Bloemen, Patrick Wils, and Pieter Degroote—for your persistence and shared vision in building PHOEBE and patience when the book kept me away from debugging.

To my colleagues Conny Aerts, Gerard van Belle, Jørgen Christensen-Dalsgaard, Remo Collet, Bernd Freytag, Bob Jantzen, and Hans-Günther Ludwig—for your help and guidance to better understand and explain pulsations, interferometers, vertical stellar atmosphere profiles, general relativity concepts, and convection.

To Frank Maloney, Simon Murphy, and Jerry Orosz—a huge debt of gratitude for reading the drafts of the book and providing me with constructive feedback.

To my mentors Tomaž Zwitter, Alojz Kodre, Bob Wilson, and Ed Guinan—you shaped me into the scientist I am today. You taught me to be independent but to rely on fellow scientists for criticism, to be ambitious but to keep a modest attitude, to be perseverant but to know when to take a break.

To the National Science Foundation (NSF) and National Aeronautics and Space Agency (NASA)—for funding the discoveries that resulted in part in this book, most notably NSF no. 1517474 and NASA 17-ADAP17-0068.

To Paula Szkody—for making me believe that I can do this.

To Leigh Jenkins, Poppy Emerson, Dan Heatley, Gus Muench, and the rest of the IOP Publishing team—for all the help, technical support, and patience with the missed deadlines.

Author biography

Andrej Prša

Andrej Prša is a Slovenian astrophysicist working in the United States. He obtained his Bachelor's Degree in Astrophysics from the University of Ljubljana in 2001 and was awarded a Preeren prize for his thesis on modeling eclipsing binary stars. He continued his graduate studies at the University of Ljubljana under the advisorship of Dr. Tomaž Zwitter and earned his PhD in Physics in 2005. The principal result of his doctoral thesis was PHOEBE (PHysics Of Eclipsing BinariEs), a modeling suite for binary stars that has since become one of the standards in the binary star community. He continued his academic career as a postdoctoral research fellow at Villanova University in Pennsylvania, USA, working with Dr. Edward Guinan on artificial intelligence approaches to modeling large swaths of binary star data coming from modern missions and surveys. Andrej participated in the operational development of the *Gaia* mission (most notably in the classification of variability within the Coordinating Unit 7), he was a member of CoRoT's Binary Star Working Group, and he chaired the *Kepler* Eclipsing Binary Working Group. The last commitment resulted in the most complete catalog of eclipsing binary stars in *Kepler*'s 105 deg^2 field of view. Andrej continues to lead the development of PHOEBE, which has grown from a small open-source program to a full-blown community project developed by an international group of scientists with roots in physics, astronomy, statistics, mathematics, and big data science. He also cochairs the Large Synoptic Survey Telescope (LSST) Interacting Binary Working Group with Dr. Paula Szkody, working toward optimizing the observing strategy to maximize the yield for binary star science. Andrej has served as president of the International Astronomical Union (IAU) Commission G1: Binary and Multiple Stellar Systems for the 2015–2018 triennium and is elected vice president of the IAU Division G: Stellar Astrophysics for the 2018–2021 triennium. He is slated to take over the presidency of Division G in 2021–2024. Andrej's research has been externally funded by NASA, NSF, and other agencies uninterruptedly since 2006; he has been the PI on 16 sponsored projects, and he received the Veritas Award for outstanding scholarship in 2018. The common thread for all his projects has been the physics of binary and multiple stellar systems. Andrej has advised and sponsored over 30 undergraduate students, graduate students, and postdoctoral research fellows; he is currently leading a vibrant research group in computational astrophysics at Villanova University. His time away from office is dedicated to his family and friends, to music and arts, and to SCUBA diving, which in many aspects Andrej considers his alternate career. He has yet to understand why anyone pays him to do what—frankly—he would do for free.

Chapter 1

Introduction to the Field

Let us start with the good news: binary star systems are as common as single stars in the universe. A review paper by Abt (1983) and subsequent studies by Raghavan et al. (2010), Duchêne & Kraus (2013), Moe & Di Stefano (2017), and many others have established that the majority of all stars are in binary or multiple stellar systems. This fact alone makes binaries appealing and worthy of studying.

1.1 Why Binaries Are Important

Binary stars prove to be the ideal test beds for a number of astrophysical studies. As they represent a significant fraction of the celestial archipelagos, they are found throughout our galaxy, in open and globular clusters, in the halo, and in other galaxies. There have been numerous studies based on binary stars, and new studies are emerging constantly. We outline below some general properties of binary stars along with the prominent references that are crucial to the field.

- The only practical way to obtain stellar masses of individual stars without external calibration (i.e., color–magnitude diagrams or mass–luminosity relations) is from binary systems. See, e.g., Martin et al. (1998) for an assembled catalog of stellar masses of astrometric binaries.
- Stellar evolution postulates the exact coevality (equal age) of both binary system components. This is of particular importance for detached binaries, because the evolution of both components is asserted to be decoupled, enabling us to study the dependence of evolutionary stages on stellar mass and luminosity. It is also important for close binaries, where the interaction influence (presence of mass transfer, circumstellar clouds) on evolutionary isochrones may be studied. See, e.g., Tohline (2002) for a review of theoretical models of binary system genesis.
- For close binaries, stellar deformation due to rotation and tidal interaction influences evolutionary pathways. See Hurley et al. (2002) for the investigation of tidal friction on binary evolution.

doi:10.1088/978-0-7503-1287-5ch1

- Binarity is not reserved for any particular type of stars; all kinds of stellar, substellar, variable, and peculiar components can be found in binaries and thus be thoroughly studied. See, e.g., Verbunt (1993) for a review of X-ray binaries, Mathieu (1994) for pre-main-sequence binaries, Taam & Sandquist (2000) for massive binaries with common envelopes, Pinfield et al. (2005) for binarity in brown dwarf systems, etc.
- Binary stars are ideal distance estimators, as absolute magnitudes of the components may be readily obtained from their luminosities. See, e.g., Munari et al. (2004) as an example of determining the distance to the Pleiades and Ribas et al. (2004) for the distance to M31 using binary stars; Guinan et al. (2004) demonstrates how to calibrate the cosmic distance scale to the Local Group using binaries.

1.2 Why Eclipsing Binaries Are Even More Important

Because of their geometrical orientation, eclipsing binaries exhibit specific features in the observed data that are easily recognized: as part of the light is blocked when one component passes in front of the other, the observed flux decreases. Such a time-dependent change in flux enables us to further constrain physical parameters of a binary system: all binary features pertain to eclipsing binaries, with eclipses as a bonus. Eclipses can, in principle, be used to obtain the following information:

- in addition to stellar masses, eclipsing binaries directly yield stellar radii (Torres et al. 2010); other methods such as interferometry or stellar occultation are much less practical because of their sensitivity to distance and unique point in time, respectively;
- the shape of the eclipses constrains the inclination;
- the duration of the eclipses and their relative position constrains orbital eccentricity, argument of periastron, and the sum of the radii of both components in units of their separation;
- the ratio of depths of the two eclipses indicates the degree of surface coverage and the relation between individual surface temperatures of both components;
- the ratio of in-eclipse to out-of-eclipse levels in different passbands constrains the amount of interstellar extinction;
- the timing of eclipses constrains the orbital period and, if significant, a period change rate and/or stellar multiplicity;
- second-order effects during eclipses constrain surface brightness variations (gravity darkening, limb darkening, reflection, spots, circumstellar matter).

These points hold only in principle and may not always yield the attributed information, depending on morphology and individual properties of the given eclipsing binary type. We will discuss these and other considerations in detail in the following chapters.

References

Abt, H. A. 1983, ARA&A, 21, 343

Duchêne, G., & Kraus, A. 2013, ARA&A, 51, 269
Guinan, E. F., Ribas, I., & Fitzpatrick, E. L. 2004, in ASP Conf. Ser. 310, IAU Colloq. 193, Variable Stars in the Local Group, ed. D. W. Kurtz, & K. R. Pollard (San Francisco, CA: ASP), 363
Hurley, J. R., Tout, C. A., & Pols, O. R. 2002, MNRAS, 329, 897
Martin, C., Mignard, F., Hartkopf, W. I., & McAlister, H. A. 1998, A&AS, 133, 149
Mathieu, R. D. 1994, ARA&A, 32, 465
Moe, M., & Di Stefano, R. 2017, ApJS, 230, 15
Munari, U., Dallaporta, S., Siviero, A., et al. 2004, A&A, 418, L31
Pinfield, D. J., Jones, H. R. A., & Steele, I. A. 2005, PASP, 117, 173
Raghavan, D., McAlister, H. A., Henry, T. J., et al. 2010, ApJS, 190, 1
Ribas, I., Jordi, C., Vilardell, F., Giménez, Á., & Guinan, E. F. 2004, NewAR, 48, 755
Taam, R. E., & Sandquist, E. L. 2000, ARA&A, 38, 113
Tohline, J. E. 2002, ARA&A, 40, 349
Torres, G., Andersen, J., & Giménez, A. 2010, A&ARv, 18, 67
Verbunt, F. 1993, ARA&A, 31, 93

Part I

Observations

Chapter 2

Observations and Data Acquisition

The purpose of this chapter is to establish the vocabulary and explain the basic principles behind the techniques frequently used in the field of binary and multiple stars. It is by no means comprehensive, and it does not aim to be complete; there are excellent books that tackle that goal successfully, most notably *Astronomical Photometry—Past, Present and Future* (Milone & Sterken 2011), *Introduction to Astronomical Spectroscopy* (Appenzeller 2013), *Astrometry for Astrophysics* (van Altena 2013), *Astronomical Polarimetry* (Tinbergen 2005) and *Astronomy at High Angular Resolution* (Boffin et al. 2016).

Essentially all the information we infer about distant astronomical objects comes from the analysis of light that is radiated from those objects. Because of the truly astronomical distances, there is currently no way to probe any astronomical objects outside our solar system directly. In fact, most of what we know about our own solar system we learned from emitted or reflected light. Fortunately, by "light" we do not mean only visible light, as that would be quite limiting. Observations in astronomy span all wavelengths; we employ special direct techniques such as interferometry or polarimetry, or indirect techniques like radial velocity (RV) measurements or Doppler tomography. In this chapter we delve into the basics of observation, focusing on the techniques relevant to the study of stars and stellar systems.

2.1 Timekeeping

Central to all types of observations is the convention that governs the recording of observation time. This is a nontrivial problem because the time of observation affects not only our accurate bookkeeping of when the observation was made but also the corrections that need to be applied to the observed quantities themselves, based on the relative motions of Earth and the observed object with respect to the chosen coordinate system. For example, if we venture to measure the absolute RV of an object to an accuracy of 1 m s^{-1} or better, it is intuitively clear that it is paramount to know accurately the absolute velocity of Earth. In addition, modern

doi:10.1088/978-0-7503-1287-5ch2

observations reach both precision and accuracy that require a proper treatment of a number of subtle relativistic effects, particularly in the *Kepler* era of photometry, HARPS era of spectroscopy, and *Gaia* era of astrometry.

2.1.1 Heliocentric Julian Date

Civil time that we use in our daily lives is not practical to time astronomical events; instead, a common practice is to use a Julian date (JD). Julian date counts the number of mean solar days that have passed since the mean Greenwich noon on 4713 BC January 1. To correct for Earth's motion, the center of gravity of the solar system is used as a reference point: this corrects for the fact that a star on the far side of the Sun requires nearly 16 more minutes for its light to reach Earth than when Earth is on the same side of the Sun as that star, approximately 6 months later. For stellar observations where timing to a minute or second is important, it is essential to apply this heliocentric correction: *heliocentric* Julian date (HJD) is the Julian date computed at the center of the Sun and can be found using the following equation (Smart & Green 1977):

$$\begin{aligned} \mathrm{HJD} = \mathrm{JD}\ (\mathrm{Geoc.}) + TR\,[&\cos\lambda_{\odot}\cos\alpha\cos\delta \\ &+ \sin\lambda_{\odot}(\sin\epsilon\sin\delta + \cos\epsilon\cos\delta\sin\alpha)], \end{aligned} \tag{2.1}$$

where $T = 499$ s is the light-travel time for 1 au, $R = 1$ au is the Sun–Earth distance, $\lambda_{\odot}$ is the ecliptic longitude of the Sun, α and δ are the star's right ascension and declination, respectively, and $\epsilon = 23.45°$ is the angle between the ecliptic and the equator. Equation (2.1) and values of the above given constants are approximate; a rigorous heliocentric correction is described, for example, in Eastman et al. (2010) and is implemented in many online calculators[1] and software packages such as IRAF (Tody 1986). The up-to-date coefficients are given in the yearly Nautical Almanacs.

Using HJD corrections is also very important for RVs, since the contribution of Earth's motion around the Sun must be properly subtracted. Heliocentric correction might be sufficient for most binary star modeling requirements, but it does not suffice for, say, pulsar timing (Hobbs & Dai 2017). In such cases it is necessary to apply the barycentric correction: *barycentric* Julian date is the Julian date referenced to the center of gravity of the solar system. Such correction is computed numerically for each point in time owing to the significant influence of many celestial bodies on Earth's motion.

2.1.2 Barycentric Celestial Reference System

One of the tasks of the International Astronomical Union (IAU) is to serve as a standards keeper. In 2000, the IAU passed resolution B1.3, which introduces the Barycentric Celestial Reference System (BCRS; Soffel et al. 2003). This reference spacetime system is described by a metric tensor $g_{\alpha\beta}$ that uniquely defines the position of a spacetime event described by four coordinates, $r^{\alpha} = (r^0, r^i) \equiv (r^0, r^1, r^2, r^3)$,

[1] I.e., http://astroutils.astronomy.ohio-state.edu/time/utc2bjd.html

where r^0 refers to the time coordinate and r^i refers to the space coordinates. Here we adopted the usual Einstein notation that uses Greek letter indices for all four spacetime components, Latin letter indices for the three space components, and where summation over repeated indices is implied. The spacetime distance differential, i.e., the distance between the events r^α and $r^\alpha + dr^\alpha$, is then given by:

$$\begin{aligned} ds^2 &= g_{\alpha\beta}(r^0, r^i) dr^\alpha dr^\beta \\ &= g_{00}(c\, dt)^2 + 2g_{0i}(c\, dt)(dr^i) + g_{ij} dr^i\, dr^j, \end{aligned} \tag{2.2}$$

where we explicitly wrote $r^0 \equiv ct$. The metric tensor then prescribes the equations of motion and light propagation and provides transformations between local reference frames, for example, the change in dt and $\boldsymbol{dr}$ from the observed object's frame to the observer's frame. Keep in mind that, according to the principle of covariance, *all* reference systems that span the same spacetime are equivalent, i.e., that the description of spacetime events can well be different, but they all describe the same reality. Thus, there is always a transformation from one reference system to another, and the ones that are best suited are those that reduce the equations and transformations to the simplest possible form. Finding such a form is the whole point of gauge fixing. The BCRS metric is given by (Soffel et al. 2003):

$$\begin{aligned} g_{00} &= -1 + \frac{2\Phi}{c^2} - \frac{2\Phi^2}{c^4} + \mathcal{O}(c^{-5}), \\ g_{0i} &= -\frac{4}{c^3}\Phi^i + \mathcal{O}(c^{-5}), \\ g_{ij} &= \delta_{ij}\left(1 + \frac{2\Phi}{c^2}\right) + \mathcal{O}(c^{-4}), \end{aligned} \tag{2.3}$$

where Φ and Φ^i are the scalar and vector potentials that relativistically generalize the Newtonian gravitational potential, and $\mathcal{O}()$ gives the power of c for which the terms containing that and higher powers are neglected. The relativistic generalization of the gravitational field is necessary because, in addition to the mass density, additional sources contribute to the gravitational field: kinetic and gravitational energy, deformation energy, rotational potential energy, etc. These additional sources contribute to the gravitational vector potential just as electric currents contribute to the electromagnetic vector potential. To further understand what Φ and Φ^i mean, consider Einstein's field equations under this metric:

$$\begin{aligned} \left(-\frac{1}{c^2}\frac{\partial^2}{\partial t^2} + \nabla^2\right)\Phi &= -4\pi G\sigma + \mathcal{O}(c^{-4}), \\ \nabla^2\Phi^i &= -4\pi G\sigma^i + \mathcal{O}(c^{-2}). \end{aligned} \tag{2.4}$$

The symbols σ and σ^i denote mass density and mass current density, respectively; thus, they generalize density ρ in the Poissonian equation $\nabla^2 U = -4\pi G\rho$ that describes the Newtonian gravitational field U. The solutions of these two equations under the boundary condition $\lim_{r\to\infty} g_{\alpha\beta} = \text{diag}(-1, 1, 1, 1)$ that imposes Euclidean flatness far from the solar system are:

$$\Phi(t, \boldsymbol{r}) = G\int \frac{\sigma(t, \boldsymbol{r})}{|\boldsymbol{r} - \boldsymbol{r}'|} d^3\boldsymbol{r}' + \frac{G}{2c^2}\frac{\partial^2}{\partial t^2}\int \sigma(t, \boldsymbol{r})|\boldsymbol{r} - \boldsymbol{r}'| d^3\boldsymbol{r},$$
$$\Phi^i(t, \boldsymbol{r}) = G\int \frac{\sigma^i(t, \boldsymbol{r})}{|\boldsymbol{r} - \boldsymbol{r}_i|} d^3\boldsymbol{r}'. \tag{2.5}$$

Thus, these are strongly reminiscent of the Newtonian gravitational potential, with the addition of the relativistic terms.

The BCRS is clearly suitable for describing motions and events in our solar system, but what about beyond? Two additional considerations are built into the metric that render the BCRS applicable for stars as well: the already-mentioned asymptotically Euclidean (flat) geometry at large distances from the Sun, and the fixed axial directions with respect to faraway extragalactic objects. The latter leaves the choice of alignment free, and the usual practice used, for example, by *Hipparcos* is to align the axes with right ascension and declination (Lindegren & Dravins 2003).

Now let us denote the time of light emission at the source as t_s and the time of detection at the observer's location as $t_{\rm obs}$. If we neglected the curvature of spacetime, the difference between the two would be simply:

$$t_{\rm obs} - t_s = \frac{1}{c}|\boldsymbol{r}_s - \boldsymbol{r}_{\rm obs}|, \tag{2.6}$$

where $|\boldsymbol{r}_s - \boldsymbol{r}_{\rm obs}|$ is the distance between the source and the observer. Because the spacetime is curved, we need to correct the above equation for the relativistic delay along the geodesic (non-Euclidean path) along which the light propagates. Let us formally denote this correction as $\gamma(t_s, \boldsymbol{r}_s, t_{\rm obs}, \boldsymbol{r}_{\rm obs})$; we can then write:

$$t_{\rm obs} - t_s = \frac{1}{c}|\boldsymbol{r}_s - \boldsymbol{r}_{\rm obs}| + \gamma(t_s, \boldsymbol{r}_s, t_{\rm obs}, \boldsymbol{r}_{\rm obs}). \tag{2.7}$$

The $\gamma(t_s, \boldsymbol{r}_s, t_{\rm obs}, \boldsymbol{r}_{\rm obs})$ term is determined by the BCRS metric tensor $g_{\alpha\beta}$. In other words, by using the $g_{\alpha\beta}$ metric, we account for all relativistic effects automatically. Let us assume for a moment that mass density is the only source of gravitational field (i.e., that $\sigma^i \equiv 0$, representing the Newtonian potential $\Phi(t, \boldsymbol{r}) = G\sum_i M_i/|\boldsymbol{r}_i|$) and that we are interested in terms up to $\mathcal{O}(c^{-2})$. The spacetime distance differential (Equation (2.2)) is then given by:

$$ds^2 = -\left(1 - \frac{2\Phi}{c^2}\right)dt^2 + \left(1 + \frac{2\Phi}{c^2}\right)dr^2 = -d\tau^2 + dl^2, \tag{2.8}$$

where we explicitly wrote the temporal and spatial differentials $d\tau$ and dl, respectively. The relationship between the actual speed of light ($dl/d\tau$) and the *coordinate* speed of light ($|dr/dt|$) is then:

$$\begin{aligned} c = \frac{dl}{d\tau} &= \frac{(1 + 2\Phi/c^2)^{1/2}}{(1 - 2\Phi/c^2)^{1/2}} \left| \frac{dr}{dt} \right| \\ &\approx \left(1 + \frac{2\Phi}{c^2}\right) \left| \frac{dr}{dt} \right| \\ \Rightarrow \left| \frac{dr}{dt} \right| &\approx c\left(1 + \frac{2\Phi}{c^2}\right)^{-1} \approx c\left(1 - \frac{2\Phi}{c^2}\right). \end{aligned} \tag{2.9}$$

Given that this is the speed of light in the BCRS, the time delay per length differential dl is then:

$$\frac{d(\Delta t)}{dl} = \frac{1}{c} - \frac{1}{|dr/dt|} \approx \frac{1}{c}\left[1 - \left(1 + \frac{2\Phi}{c^2}\right)\right] \approx -\frac{2\Phi}{c^3}. \tag{2.10}$$

We now need to integrate this expression along the line of sight between the source ($\boldsymbol{r}_s$) and the detector ($\boldsymbol{r}_{\mathrm{obs}}$), which is a straight line in the BCRS. That will yield a relativistic correction for the cumulative time delay, i.e., the term that we denoted as γ in Equation (2.7):

$$\gamma(t_s, \boldsymbol{r}_s, t_{\mathrm{obs}}, \boldsymbol{r}_{\mathrm{obs}}) = \sum_i \frac{2GM_i}{c^3} \ln\left(\frac{\hat{\boldsymbol{l}} \cdot (\boldsymbol{r}_s - \boldsymbol{r}_i) + |\boldsymbol{r}_s - \boldsymbol{r}_i|}{\hat{\boldsymbol{l}} \cdot (\boldsymbol{r}_{\mathrm{obs}} - \boldsymbol{r}_i) + |\boldsymbol{r}_{\mathrm{obs}} - \boldsymbol{r}_i|}\right). \tag{2.11}$$

Here the index i goes over all mass sources between the object and the observer, and we denoted a unit vector along the line of sight as $\hat{\boldsymbol{l}} \equiv (\boldsymbol{r}_s - \boldsymbol{r}_{\mathrm{obs}})/|\boldsymbol{r}_s - \boldsymbol{r}_{\mathrm{obs}}|$. In reality we rarely (if ever) know all mass sources that shape the gravitational field along the entire path. It thus proves useful to explicitly separate the local (solar system) contributions from all others. If we limit the mass sources to the solar system, Equation (2.11) can be simplified by noting that $\boldsymbol{r}_s \gg \boldsymbol{r}_i$ and that $\hat{\boldsymbol{l}} \| \boldsymbol{r}_s$:

$$\begin{aligned} \gamma(t_s, \boldsymbol{r}_s, t_{\mathrm{obs}}, \boldsymbol{r}_{\mathrm{obs}}) = {} & \frac{2G}{c^3} \sum_i M_i \ln \frac{2r_s}{A} \\ & - \frac{2G}{c^3} \sum_i M_i \ln\left(\frac{\hat{\boldsymbol{l}} \cdot (\boldsymbol{r}_{\mathrm{obs}} - \boldsymbol{r}_i) + |\boldsymbol{r}_{\mathrm{obs}} - \boldsymbol{r}_i|}{A}\right), \end{aligned} \tag{2.12}$$

where A is an arbitrary value, set by convention to 1 au: that is a typical order of magnitude of the denominator in Equation (2.11), as the Sun is by far the most important contributor. The first term in Equation (2.12) is effectively a constant for the given star, as it depends only on its distance, r_s. The second term depends on the location of the observer with respect to the barycenter of the solar system. It is now straightforward to compute a barycentric correction to the observed time t_{obs}:

$$t_{\mathrm{B}} = t_{\mathrm{obs}} + \frac{(r_s - r_t)}{c} + \frac{2G}{c^3} \sum_i M_i \ln\left(\frac{\hat{\boldsymbol{l}} \cdot (\boldsymbol{r}_{\mathrm{obs}} - \boldsymbol{r}_i) + |\boldsymbol{r}_{\mathrm{obs}} - \boldsymbol{r}_i|}{A}\right), \tag{2.13}$$

where $r_t \equiv |\boldsymbol{r}_s - \boldsymbol{r}_{\mathrm{obs}}|$ is the topocentric coordinate distance to the source. Equation (2.13) is known as the barycentric correction, and it includes all relativistic effects due to masses in the solar system. Those with a keen eye will catch that we need to know the distance r_s to the system to evaluate Equation (2.13), but (a) the distance needs to be known only to a moderate precision, (b) this is no longer an insurmountable task since *Gaia* delivered accurate distance measurements across our Galaxy, and (c) for most practical purposes the curvature terms with r_s can be neglected and the results will still be well within the measurement uncertainty (Lindegren & Dravins 2003).

Before we close our discussion on timekeeping, let us visit a question of time as witnessed by the source versus time as witnessed by the observer. As before, let us denote with t_s a time interval at the source and with $t_{\rm obs}$ a time interval at the observer's location. We start by taking the differential of Equation (2.7):

$$dt_{\rm obs} - dt_s = \frac{1}{c}\hat{\boldsymbol{l}} \cdot (\boldsymbol{r}_s - \boldsymbol{r}_{\rm obs}) + d\gamma(t_s, \boldsymbol{r}_s, t_{\rm obs}, \boldsymbol{r}_{\rm obs}). \tag{2.14}$$

Here we used $|\boldsymbol{r}_s - \boldsymbol{r}_{\rm obs}| \equiv \hat{\boldsymbol{l}} \cdot (\boldsymbol{r}_s - \boldsymbol{r}_{\rm obs})$. Next, we divide the equation by $dt_{\rm obs}$:

$$1 - \frac{dt_s}{dt_{\rm obs}} = \frac{1}{c}\left(\frac{d\boldsymbol{r}_s}{dt_{\rm obs}} - \frac{d\boldsymbol{r}_{\rm obs}}{dt_{\rm obs}}\right) + \frac{d\gamma}{dt_{\rm obs}}. \tag{2.15}$$

The two derivatives inside the parentheses are source and observer velocities in the observer's time frame, respectively; we need to switch the source velocity into the source time frame by applying the chain rule, $d\boldsymbol{r}_s/dt_{\rm obs} = (d\boldsymbol{r}_s/dt_s)(dt_s/dt_{\rm obs})$, plugging it into Equation (2.15), and expressing $dt_s/dt_{\rm obs}$:

$$\frac{dt_s}{dt_{\rm obs}} = \frac{1 + \hat{\boldsymbol{l}} \cdot \boldsymbol{v}_{\rm obs}/c - d\gamma/dt_{\rm obs}}{1 + \hat{\boldsymbol{l}} \cdot \boldsymbol{v}_s/c}. \tag{2.16}$$

For most practical purposes the term $d\gamma/dt_{\rm obs}$ is negligible (Lindegren & Dravins 2003), so the sought relationship between the time interval at the source and the time interval at the observer's location is:

$$\frac{dt_s}{dt_{\rm obs}} = \frac{1 + \hat{\boldsymbol{l}} \cdot \boldsymbol{v}_{\rm obs}/c}{1 + \hat{\boldsymbol{l}} \cdot \boldsymbol{v}_s/c}. \tag{2.17}$$

Thus, the time interval difference at the source and at the observer depends on the velocity of the source with respect to BCRS and the velocity of the observer with respect to BCRS. This enables us to properly account for the relativistic effects on measured quantities.

Having survived this digression into general relativity, those looking to easily convert to and from the BCRS will find the previously mentioned reference (Eastman et al. 2010) and the corresponding calculators very useful.

2.2 Photometry

The most accessible data acquisition method in astronomy is undoubtedly photometry. It provides a measurement of the amount of light intercepted by a detector. It differs from photography by the ability to precisely convert brightness into a physical quantity, most notably flux or magnitude. Early days of photometry witnessed extensive use of photographic plates where the determination of brightness hinged on the properties of the photographic grains. Notable efforts are underway to scan the plates and convert the photographic brightnesses to photometric values (Tsvetkov et al. 1994; Grindlay et al. 2012; Groote et al. 2014). The first electronic instrument for acquiring photometric observations was a photoelectric photomultiplier. It recorded the brightness from a single source by measuring the (amplified) energy of the photons deposited on the detector. The principal instrument for acquiring photometric observations in use today is a charge-coupled device (CCD): an array of photosensitive pixels that records the number of photons striking each pixel. CCDs took over because of their versatility, wide spectral range, and good per-pixel price value that enables large fields of view and observations of many targets simultaneously.

Photometric observations that we acquire depend on many factors: the intrinsic properties of the light source, the presence of interstellar matter that may cause reddening and extinction, scattering and extinction within our atmosphere for ground-based instruments, instrumental efficiency of the optical assembly, filter response, and detector sensitivity (quantum efficiency or photocathode radiant response). These effects need to be either corrected or accounted for. Such instrumental flux is only a part of the total (bolometric) flux emitted by the object that reaches our atmosphere:

$$F_\lambda^{\mathrm{instrumental}} = \mathcal{P}(\lambda)\, R^{\mathrm{ccd}}(\lambda)\, R^{\mathrm{optics}}(\lambda)\, R^{\mathrm{atmosphere}}(\lambda)\, F_\lambda^{\mathrm{bolometric}}, \tag{2.18}$$

where $\mathcal{P}(\lambda)$ is the filter transmission function. Variations in the atmosphere, $R^{\mathrm{atmosphere}}(\lambda)$, may be corrected by monitoring the flux variation of comparison stars that are present in the field. The remaining factors depend on the optical layout of the instrumental system and can be calibrated using standard stars in the field during data reduction by dedicated tools such as IRAF (Tody 1986). The ultimate goal is to transform this instrumental flux (or more commonly the magnitude) into a standard photometric system, i.e., Johnson (1965), Cousins (1976), Bessell (1990), or something similar. Figure 2.1 depicts filter transmission curves of the Bessell (1990) *UBVRI* series, a CCD response curve of the commercial Kodak KAF-16803 CCD chip, and atmospheric transmission at Mauna Kea for objects at zenith.

Multiple photometric observations as a function of time constitute a *light curve*. Light curves can reveal variability of the source, be it intrinsic or aspect dependent. If a variability is periodic, we can fold the observations to the phase interval [0, 1] by providing the time stamps modulo the period:

$$\Phi = \frac{(t - t_0) \bmod P_0}{P_0}, \tag{2.19}$$

where t_0 is a reference time and P_0 is the variability period. When phasing binary star light curves, a [−0.5, 0.5] interval is often preferred to center the primary minimum that by convention coincides with $\Phi = 0$. Figure 2.2 shows several examples of the

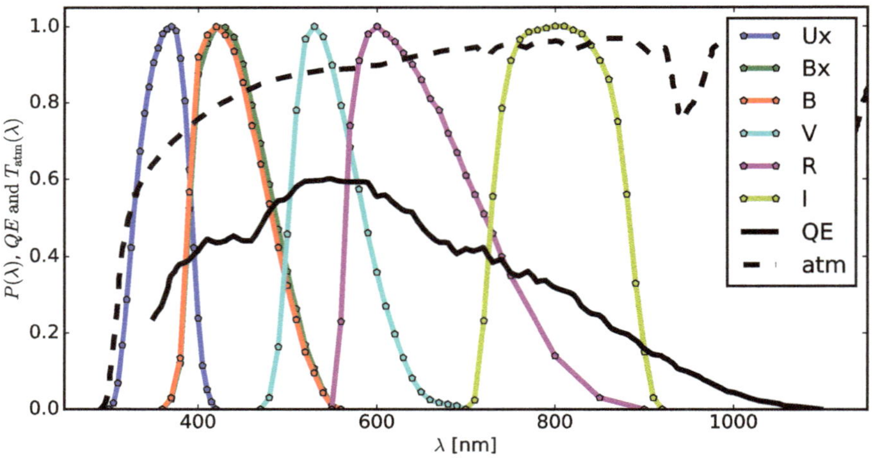

Figure 2.1. Filter transmission, CCD quantum efficiency, and atmosphere transmission curves. Filter transmissions are depicted for the Bessell (1990) filter set; quantum efficiency (QE) corresponds to Kodak's KAF-16803 CCD chip, taken from the Eastman Kodak Company (2012) specifications sheet; and atmospheric transmission is computed with a LOWTRAN VI theoretical model for the Mauna Kea location and objects at zenith (Nitschelm 1988).

photometric time-series and phased light curves of the *Kepler* eclipsing binary stars with pulsating components. The data are adopted from Kirk et al. (2016).

Photometric accuracy is typically conveyed in terms of the *signal-to-noise ratio* (S/N): the power ratio between a signal (meaningful information) and the background noise:

$$\mathrm{S/N} = \frac{P_{\mathrm{signal}}}{P_{\mathrm{noise}}}. \tag{2.20}$$

The amount of photometric noise present in the signal depends on the photon regime of the observations:

SEEING-LIMITED regime is inherent to bright sources, where the atmospheric scintillation is the dominant source of error. The noise scales *linearly* with the light level.

SHOT-LIMITED REGIME is inherent to the sources from which enough photons are acquired that photon collection is described by Poissonian statistics. The noise scales with the *square root* of the light level.

SKY-LIMITED REGIME is inherent to faint sources, where background fluctuations are of the same order of magnitude as the signal. The noise scales *linearly* with the light level.

Most photometrically attractive objects (not too bright, not too dark) fall into the shot-limited regime, which makes photometry one of the most accurate data acquisition methods in astronomy.

Astronomers often prefer to express fluxes in terms of magnitudes. The transformation to magnitudes is given by:

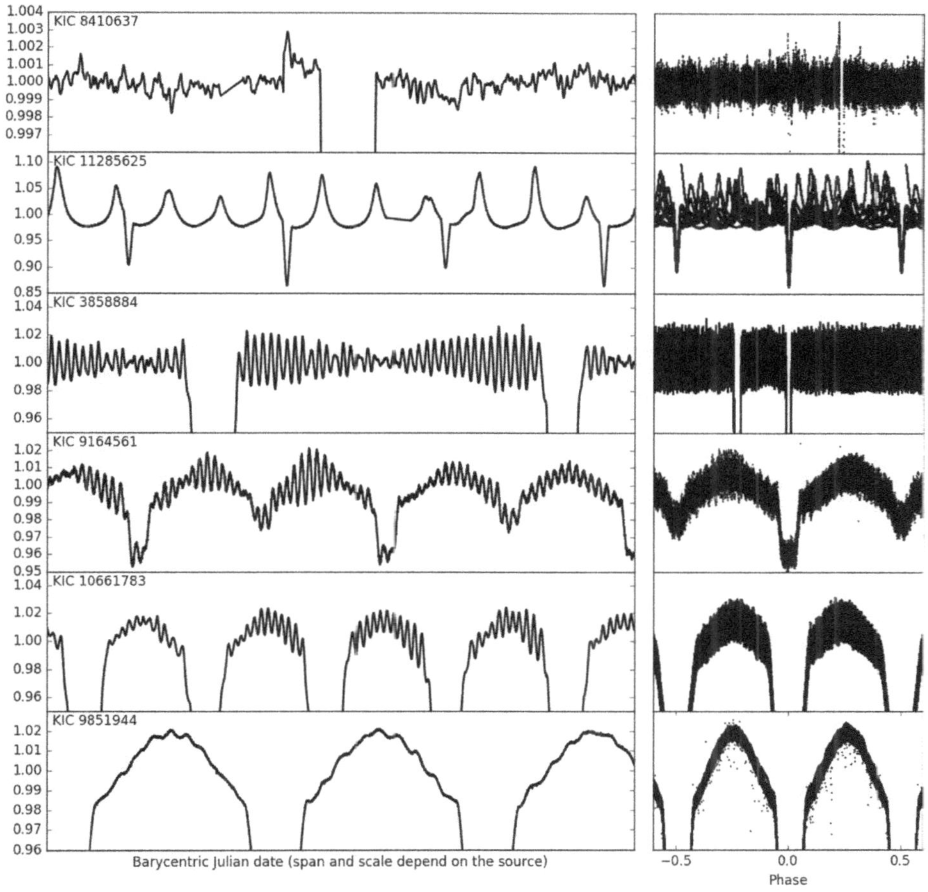

Figure 2.2. Light curves of several eclipsing binary stars with pulsating components. Left panels depict time sequences and right panels depict phase plots for each object, where time span and scale for time sequences have been chosen for each object individually to highlight the telling part of the light curve. The eclipsing binaries contain a red giant exhibiting solar-like oscillations (KIC 8410637), a γ Dor pulsator (KIC 11285625), a tidally induced heartbeat star (KIC 3858884), a pre-He-WD pulsator (KIC 9164561), a δ Sct pulsator (KIC 10661783), and a hybrid δ Sct/γ Dor pulsator (KIC 9851944). All data are from the *Kepler* mission.

$$m - m_0 = -\frac{5}{2}\log\left(\frac{F}{F_0}\right), \tag{2.21}$$

where m_0 is the magnitude zero-point that corresponds to the flux F_0. It is important to note that the errors in measured flux do not transform symmetrically to errors in magnitudes. Given a standard deviation in flux, $\pm 1\sigma_F$, we can write:

$$F^{\pm} = F \pm \sigma_F \quad \Rightarrow \quad \frac{F^{\pm}}{F} = 1 \pm \frac{\sigma_F}{F}, \tag{2.22}$$

where F is the observed photometric data point in flux units. To convert the uncertainty to magnitudes, we use Equation (2.21):

$$\sigma_m^{\pm} = -\frac{5}{2}\log\left(1 \pm \frac{\sigma_F}{F}\right). \tag{2.23}$$

The absolute values of the two deviations, σ_m^+ and σ_m^-, are not equal: σ_m^+ is always larger than σ_m^-. In practice, we assign either σ_m^+ or the average of the two to photometric data points given in magnitudes.

We will return to radiative quantities in Chapter 4 and discuss them in detail, but let us use this opportunity to stress three important points worth keeping in mind.

The first point concerns the detectors: while older detectors (such as photomultipliers) measure the *energy* of photons striking the detector, CCDs measure *photon counts*. In other words, when we say "flux," we mean either energy flux (in W m^{-3} or, more practically, W m^{-2} $Å^{-1}$) or photonic flux (in photons s^{-1} m^{-3} or photons s^{-1} m^{-2} $Å^{-1}$). It is important to distinguish between the two regimes because the response of the detector is different and it affects the computation of synthetic fluxes that we compare with observations. A detailed discussion of this point is given by Brown et al. (2016b), Bessell & Murphy (2012), and Casagrande & VandenBerg (2014).

The second point concerns the quantity being measured. We referred to either "light" or "flux" (or magnitude), but the quantity that measures the amount of energy per unit time radiated from the projected unit area toward the observer is *intensity*. For unresolved sources (i.e., sources smaller than the diffraction limit of the instrument) this makes no difference, as *all* light is coming from the source; for resolved sources, however, the amount of light corresponds to the part of the extended object we are observing, which corresponds to intensity, not flux. In practice, this affects only a handful of nearby stars (with the Sun being the most obvious), but with developing technology and larger telescopes, making this distinction is becoming increasingly important.

The third and final point concerns the exposure time. Depending on the size of the instrument and other observing conditions, photometric observations of faint objects may necessitate that photons are collected for an extended period of time. What "extended" means depends on the nature of the object, but in general, whenever the exposure time is of the order of the variability timescales present in the light of the observed object, our observations will necessarily suffer from temporal averaging. This effect, known as finite time of integration (FTI), smooths the light curves of variable objects.

2.3 Spectroscopy

Spectrographs disperse a beam of light into a wavelength-distributed spectrum. Spectra come in all shapes and forms, from prism-dispersed rainbows to high-resolution spectral line profiles, from X-rays to radio waves. Optical spectra usually describe flux (or intensity) as a function of either wavelength or frequency. The resolving power of the spectrum is defined as:

$$R = \frac{\lambda}{\Delta\lambda}, \tag{2.24}$$

where $\Delta\lambda$ is *dispersion*, i.e., the width of the smallest discernible unit of wavelength around λ. What features we can resolve in a spectrum directly depends on the resolving power of the spectrograph. At $R \sim 500$ (low-resolution spectroscopy), only basic continuum features are typically discerned; at $R \sim 5000$ (medium-resolution spectroscopy), equivalent widths of strong spectral lines may be studied; at $R \gtrsim 20,000$ or higher (high-resolution spectroscopy), narrow spectral lines may also be studied and, with increasing resolution, finer features such as line broadening or line splitting may be observed. Modern echelle spectrographs typically reach well over $R \sim 100,000$. Figure 2.3 shows a part of the reduced (wavelength-calibrated, flux-normalized) echelle spectrum near Hα of the eclipsing binary UV Leonis. The spectrum was acquired using the 1.82 m Mount Ekar telescope in Asiago, Italy.

There are four principal physical parameters that predominantly drive the features in high-resolution spectra:

EFFECTIVE TEMPERATURE (T_{eff}) of the source is the dominant parameter that influences the shape of the spectral energy distribution (SED) and the presence and strength of spectral lines.

SURFACE GRAVITY ($\log g$) of the source indicates the size of the star; it influences the shapes of spectral lines. Gravitational acceleration g is assumed to be in cgs units, which is an unfortunate relic of the past, both because it does not use MKS system of SI units and because the argument to the log function should be dimensionless. It would have been far better to refer to surface gravity as $\log g/g_0$, where $g_0 = 1$ cm s^{-2}, which would have fixed both issues, but we are stuck with this convention for the time being and remind again that cgs units are implied.

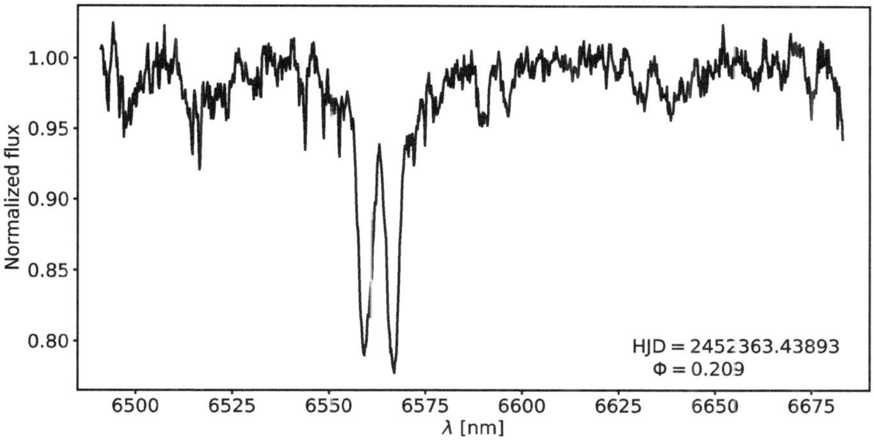

Figure 2.3. Part of the echelle spectrum of the eclipsing binary UV Leonis. The Doppler-splitting of Hα due to the binary nature of the system is clearly visible. The spectrum was acquired with the echelle spectrograph mounted on the 1.82 m Mount Ekar telescope in Asiago, Italy, and reduced by IRAF (Tody 1986).

METALLICITY ([M/H]) of the source indicates the abundance of heavy elements compared to solar abundances on a logarithmic scale; it influences the presence and strength of spectral lines of elements heavier than helium. Metallicity [M/H] = 0 corresponds to solar abundances, [M/H] = −1.0 means 10 times fewer metals, and [M/H] = 1.0 means 10 times more metals. The symbol M in [M/H] stands for all elements heavier than helium; sometimes other abundances are published, i.e., [Fe/H] for iron, [α/H] for α-particle capturing elements, etc.

ROTATIONAL VELOCITY v_{rot} of the source governs the amount of Doppler broadening of spectral lines. The actually measured quantity is the line-of-sight-projected rotational velocity $v_{rot} \sin i$, where i is orbital inclination: $i = 90°$ corresponds to edge-on systems, and $i = 0°$ corresponds to face-on systems.

Astronomers often use spectral types instead of effective temperatures: spectral type O corresponds to the hottest blue stars ($T_{eff} \sim 50,000$ K), and following in decreasing temperature are spectral types B ($T_{eff} \sim 20,000$ K), A ($T_{eff} \sim 8500$ K), F ($T_{eff} \sim 6600$ K), G ($T_{eff} \sim 5700$ K), K ($T_{eff} \sim 4600$ K), and M ($T_{eff} \sim 3200$ K). This spectral taxonomy was formally established by Pickering, but the bulk of the grunt work was done by Annie Jump Cannon in the 1890s, and the final revision was done by Cecilia Payne (1925) in her PhD dissertation. Each spectral type has subtypes that range from 0 to 9 (i.e., G0 through G9), where 0 is the hottest and 9 is the coolest subtype. Decimal subtypes are also allowed and used mostly for O-type stars, i.e., O9.5.

Closely related to spectral classification are luminosity classes that are used instead of log g: type V corresponds to main-sequence (dwarf) stars, IV to subgiants, III to normal giants, II to bright giants, and I to supergiants. This classification is established by Morgan et al. (1943). For example, the Sun is a G2V star.

Figures 2.4–2.7 depict the influence of the four listed parameters on the observed stellar spectra in the *Gaia* wavelength range (850–875 nm). The spectra are acquired with the echelle spectrograph mounted on the 1.82 m Mount Ekar telescope in Asiago, Italy.

When observations are acquired at a high resolving power and a high S/N, the spectra can reveal much more than just those four parameters. We can commonly measure the abundances of individual elements and microturbulent velocity, infer on the local conditions from line broadening, determine dynamical properties from Doppler shifts of spectral lines, and, at a very high R and S/N, we can even perform Doppler tomography: the study of surface features by time-variable spectral line profiles. We discuss the spectroscopic techniques especially important for the study of binary stars below.

2.3.1 Radial Velocities

The Doppler effect—the change in the wavelength due to the kinematic motion of the radiating body toward or away from the observer —causes excursions of spectral

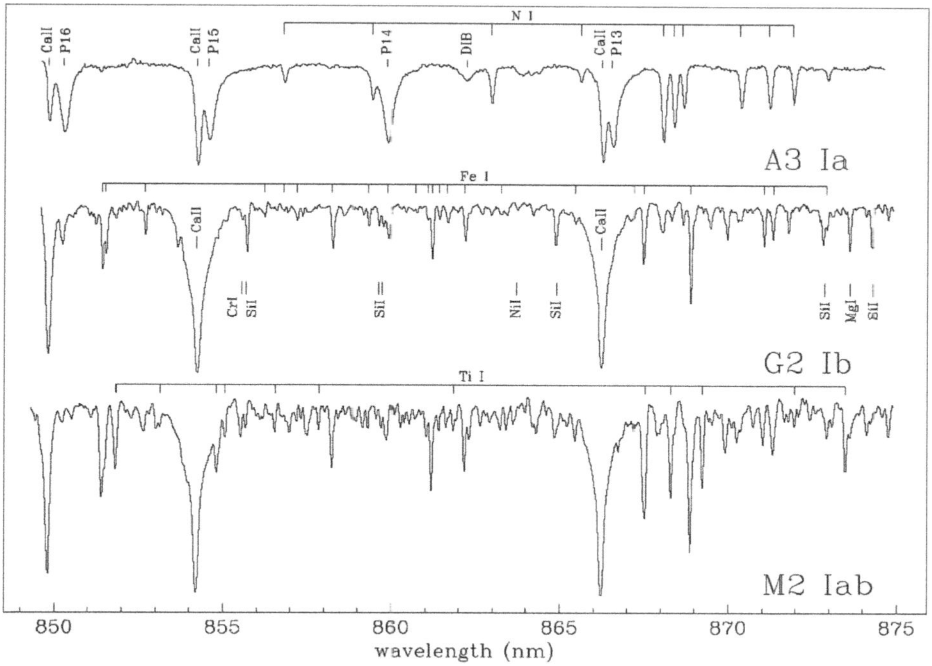

Figure 2.4. Influence of effective temperature (spectral type) on the stellar spectra. Adapted from Munari and Tomasella (1999). Credit: U. Munari and L. Tomasella, A&A. Suppl. Ser., 137, 521-528, 1999, reproduced with permission © ESO.

lines from their rest wavelengths. In binary (and multiple) systems, these excursions are caused by the radial component of the orbital velocity with respect to the barycenter of the system. When we detect the wavelength shifts in the binary system spectra, we refer to those systems as *spectroscopic* binary stars. Depending on the light contribution from each component, spectroscopic binaries can either be single-lined (SB1), where most of the light originates from one component, or double-lined (SB2), where the contributions from both components are significant. This designation can be easily generalized further for multiple systems (SB3, SB4, …). As these wavelength excursions provide deep insight into the dynamics and the absolute scale of the binary/multiple systems, measuring them accurately is a major task for stellar spectroscopy.

Consider a spectrum of V505 Persei, depicted in Figure 2.8 (Munari et al. 2001). Extracting RV data from spectra boils down to measuring the Doppler shifts accurately. This may seem straightforward, but it is not. People have historically fitted Gaussian profiles to individual spectral lines and deduced RVs from the obtained Gaussian means, yet this approach is prone to significant inaccuracies, especially in the vicinity of eclipses where the lines are not resolved (Rossiter 1924). Fitting two Gaussians simultaneously improves the accuracy, and this is the technique used, for example, in IRAF (Tody 1986). Other specialized techniques have been developed for RV measurements. One such example is TODCOR (TwO-

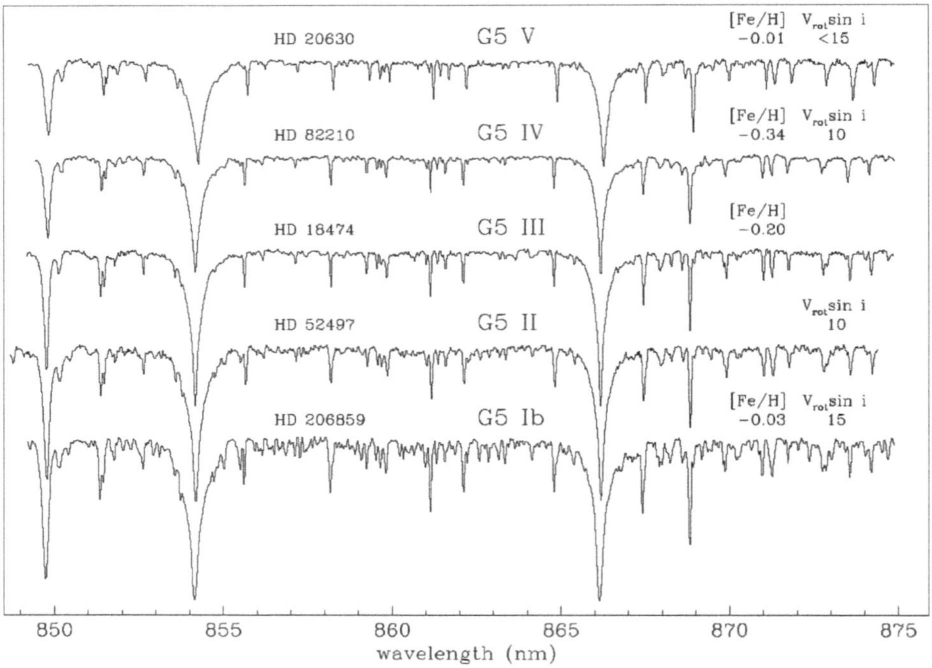

Figure 2.5. Influence of surface gravity (luminosity class) on the stellar spectra. Adapted from Munari and Tomasella (1999). Credit: U. Munari and L. Tomasella, A&A. Suppl. Ser., 137, 521-528, 1999, reproduced with permission © ESO.

Dimensional Cross-cORrelation; Zucker & Mazeh 1994) and its generalized versions TRICOR (Zucker et al. 1995) for SB3 systems and TRIMOR (Mazeh et al. 2009) for multiorder spectra of SB3 systems. The basic consideration of this technique is cross-correlation of template spectra with the observed spectrum that yields Doppler shifts that maximize the composite overlap between the templates and the observed spectrum. Another such example is *disentangling* (Simon & Sturm 1994; Hadrava 2009). This method relies on decomposing the light from the system into individual object contributions by simultaneously determining the underlying spectrum and the orbital properties of the system. Finally, the use of broadening functions (Rucinski 1992) has been proven meritorious for resolved components in the spectra; the main difference is that the method derives the broadening kernel instead of a deconvolved spectrum, i.e., it finds the parameters that describe the entire line profile (including shifts) rather than the entire component spectra.

There is a fine line between the measured and kinematic RV. Kinematic RV corresponds to the geometric projection of the orbital velocity along the line of sight; as such, it depends on the chosen spacetime metric (see Section 2.1). In contrast, the observed RV is a *measure*, i.e., it is a quantity measured within the barycentric frame of the solar system, and it does *not* depend on the spacetime metric—i.e., it cannot be converted to the kinematic RV without further assumptions. When we say

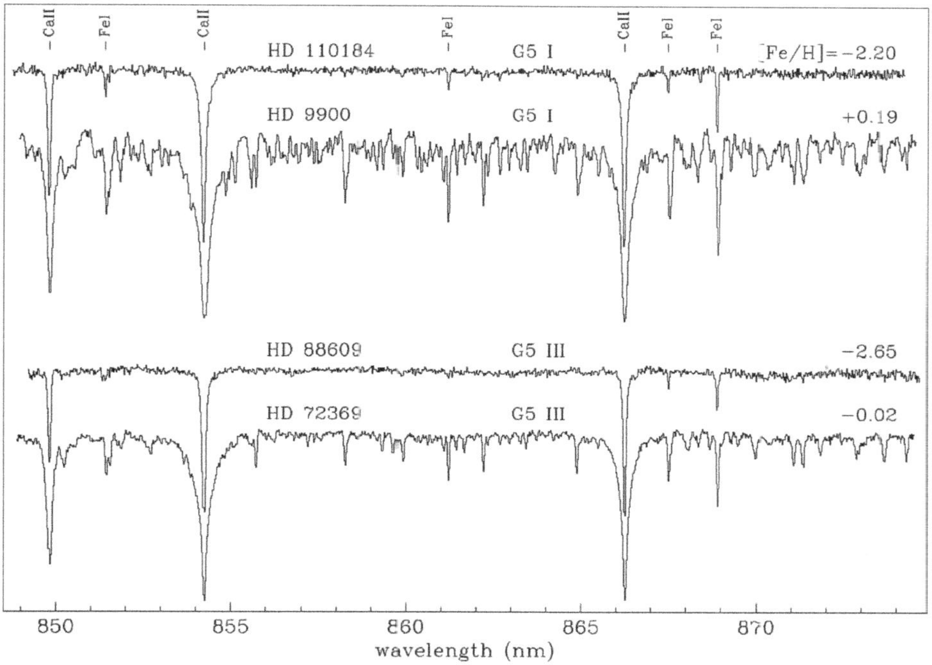

Figure 2.6. Influence of metallicity on the stellar spectra. Adapted from Marrese et al. (2003). Credit: P. M. Marrese; F. Boschi; U. Munari, A&A, 406, 995, 2003, reproduced with permission © ESO.

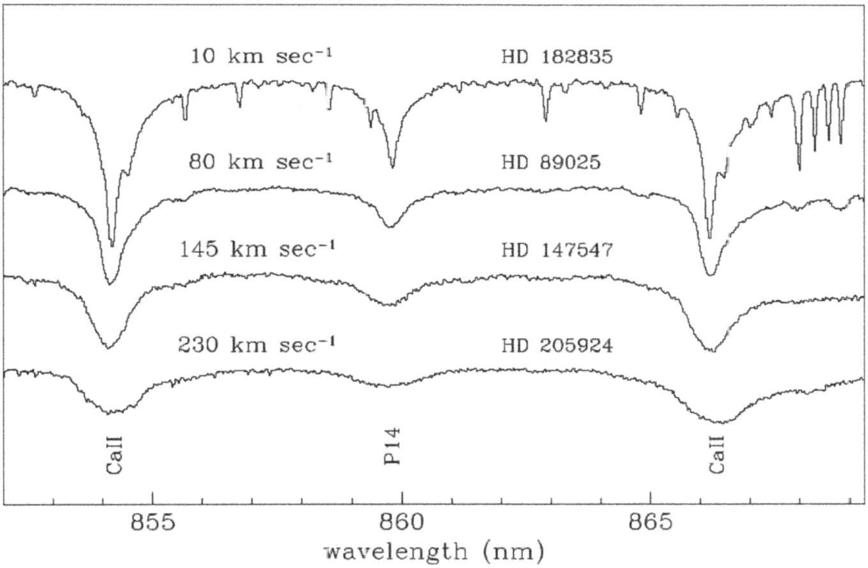

Figure 2.7. Influence of rotational velocity on the stellar spectra. Adapted from Munari (1999).

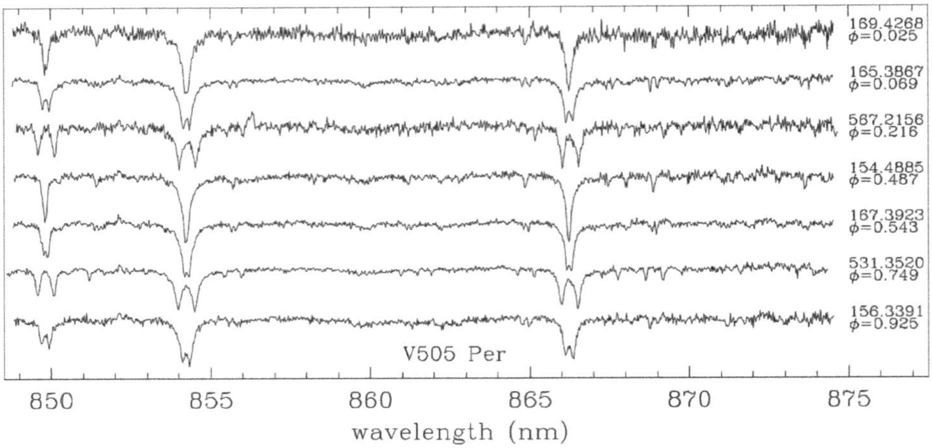

Figure 2.8. Spectra of V505 Per, a double-lined spectroscopic binary (SB2) with similar components. The time stamps (HJD –2,451,000) and phases for each spectrum are listed on the right. Spectral lines are split because of the orbital motion of the two stars about the center of mass. Adopted from Munari et al. (2001). Credit: U. Munari et al., A&A, 378, 477, 2001, reproduced with permission © ESO.

"observed RV," we really mean the *barycentric RV measure* (Lindegren & Dravins 2003).

From the measured RVs we can assemble an RV curve, as a function of either time or phase. The RV curve is fundamentally different from a light curve: while a light curve provides a directly observed quantity (flux), the RV curve provides a *derived* quantity (RVs). Strictly speaking, comparing the barycentric RV measure (derived from spectra) and kinematic RVs computed from orbital parameters is not directly possible. The appropriate approach to compare the two would be to synthesize a combined SED of the system and apply the same measurement technique (fitting a Gaussian, TODCOR, disentangling, etc.) to obtain a synthetic RV measure. While this is certainly a viable approach, it is not practical and, in most circumstances, the precision of the RV measures is such that the differences between kinematic and measured RVs are buried in noise. Thus, we will adopt a lax terminology from now on and refer to barycentric RV measures simply as RVs, and we will stress the distinction only when it is relevant for the discussion. The RV curves can thus *informally* be considered as analogs to light curves. Using light and RV curves of appropriate eclipsing binary systems simultaneously allows us to determine the fundamental stellar parameters (masses, radii, luminosities, and temperatures) of the binary/multiple system components. Photometric light curves provide us with the relative proportions of the system, and RV curves determine the absolute scale of the system. Figure 2.9 depicts an example of the extracted RVs for KIC 4150611, a quintuple SB3 system discovered by the *Kepler* satellite (Prša et al. 2011). The observations are acquired by the SAO Tillinghast Reflector Echelle Spectrograph (TRES) mounted on the 1.5 Fred Whipple telescope on Mount Hopkins in Arizona.

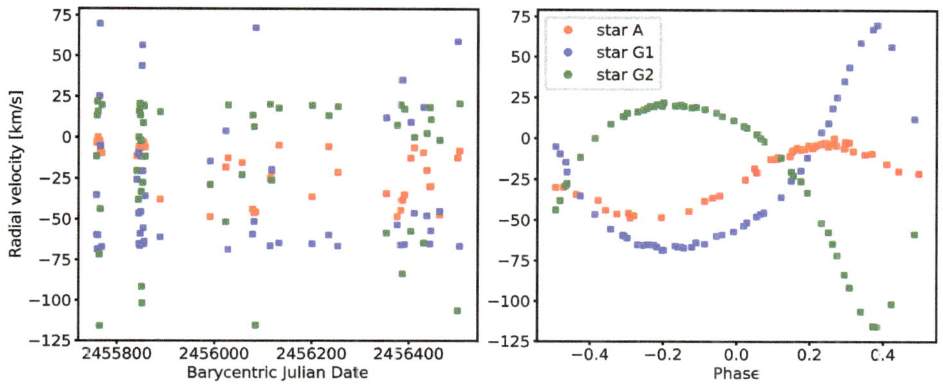

Figure 2.9. RV curve of the quintuple SB3 system KIC 4150611. The system consists of a rapidly rotating A-type star, orbited by a pair of M dwarfs (invisible in the spectra), and this triple in turn is orbited by a pair of G-type dwarfs. Left: RV curve in time space. Right: RV curve in phase space, where the A-star RV curve and the G1/G2-star RV curves are phased with their respective orbital periods, 94.2 and 8.65 days. The data are folded to the [−0.5, 0.5] phase interval.

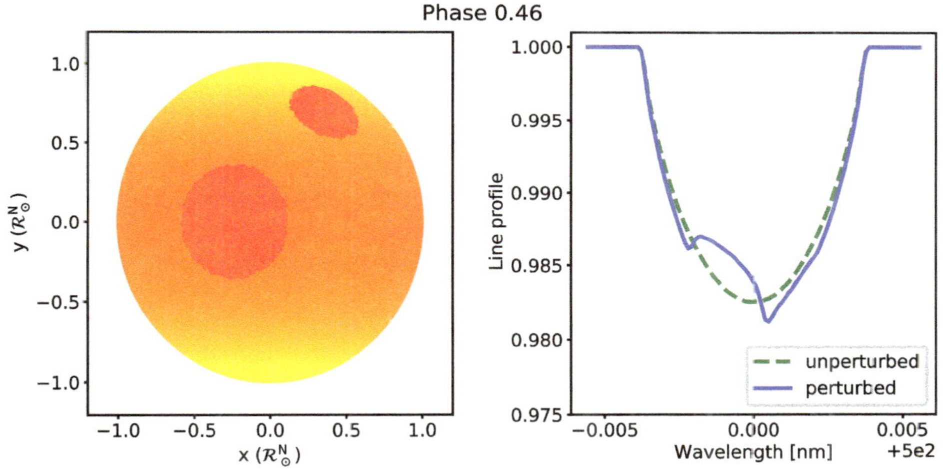

Figure 2.10. Influence of surface prominences on the spectral line profile. The left panel depicts a star with two exaggerated spots. The larger spot angular radius is 20°, and its temperature is 80% that of the underlying photosphere. The smaller spot angular radius is 13°, and its temperature is 85% that of the underlying photosphere. The right panel depicts a line profile at that particular phase, as seen from Earth. An animated schematic is available in the interactive version of the book or from the book homepage, available online at https://doi.org/10.1088/978-0-7503-1287-5.

2.3.2 Doppler Tomography

When very high resolution spectra are available (R = 50,000 or higher) and a reasonable phase coverage is acquired, we can obtain accurate measurements of the spectral line profiles. The idea is to reverse-map distortions of spectral lines onto particular features on the stellar surface. Figure 2.10 depicts a schematic of a star with two exaggerated spots, one 20° in radius and 20% cooler than the underlying photosphere, and the other 13° in radius and 15% cooler than the underlying photosphere. The corresponding line profile is shown on the right. The change in line profile is due to the suppression of flux by the spots, and it depends on the phase at which we see the star. Doppler tomography converts the temporal dependence of the line profile into the location of surface features that suppress or enhance the flux. The method is convenient for rapidly rotating single stars and, to our benefit, most types of close binary stars.

Doppler tomography (or Doppler imaging) was first proposed for astronomical application by Deutsch (1957); with improved mathematical and numerical techniques, Doppler tomography quickly gained traction in many astronomical problems: for the analysis of rapid stellar rotation (Vogt et al. 1987), magnetic Ap stars (Hatzes 1988), chemically peculiar stars (Rice et al. 1989), abundance inhomogeneities (Piskunov & Wehlau 1990), surface spots on late-type stars (Vogt & Penrod 1983; Piskunov et al. 1990), dwarf novae (Marsh et al. 1990), eclipsing binaries (Vincent et al. 1993), nonradial pulsation modes (Hao 1998), contact binaries (Hendry & Mochnacki 2000), differential rotation (Barnes et al. 2000), gravity darkening (Kovári et al. 2004), accretion disks and streams (Agafonov et al. 2009), brown dwarfs and exoplanets (Crossfield 2014), and planetary disks (Manser et al. 2016).

Doppler tomography of eclipsing binaries aims to detect and map the presence of stellar spots, center-to-limb intensity variation (limb darkening), rotational and tidal deformation (gravity darkening) effects, and other nonuniform structures on stellar surfaces of both components. Doppler shift of the distortion is determined by the position of the mapping feature on the surface (stellar latitude and longitude). Most tomography methods use the method of maximum entropy (Vogt et al. 1987) to solve the inverse problem. Figure 2.11 shows an example of a Doppler tomography reconstruction of the interacting binary star σ Coronae Borealis; the results are adopted from Strassmeier and Rice (2003). The top panel corresponds to the primary star, and the bottom panel corresponds to the secondary star. The results indicate that Doppler tomography of eclipsing binaries performs better than for single rapidly rotating stars, because the orbital movement and eclipses break the hemisphere degeneracy that the feature is on.

2.4 Interferometry

The limit on resolving power of any telescope is ultimately imposed by diffraction. The maximum resolving angle of an instrument with a linear aperture can be easily derived from first principles. Consider the incoming wavefront depicted on Figure 2.12. The linear wavefront can be thought of as an ensemble of many adjacent sources of concentric waves (this is known as the Huygens principle). As the wavefront hits the optical barrier, the light that passes through the linear aperture is then diffracted into the geometrical shadow precisely because of this principle. Most of the light will propagate straight through, and that is why the point on the screen that is perpendicular to the direction of propagation will be the brightest. As we move away from the center, the intensity of light will decrease because the distance

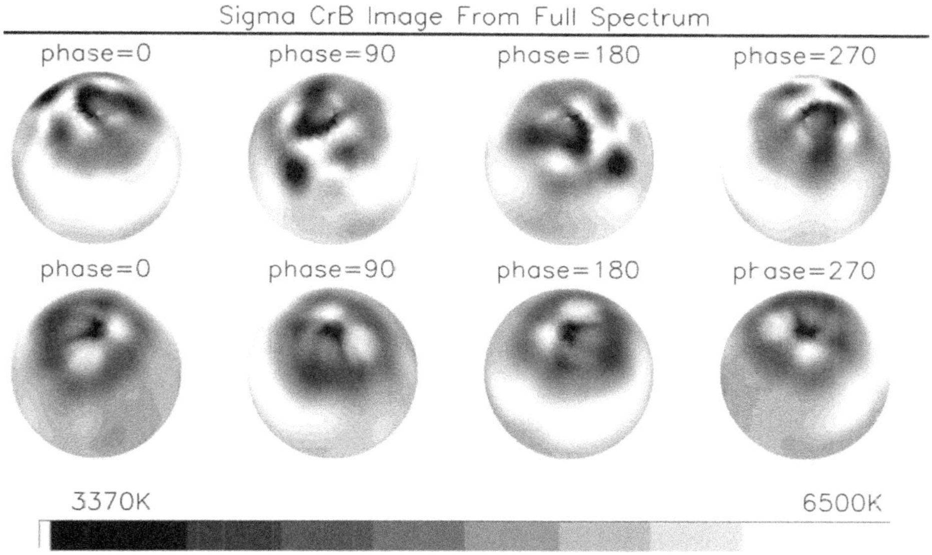

Figure 2.11. Doppler image of the components of the interacting binary σ CrB. The reconstruction was done by the simultaneous inversion of the 639–645 nm region and fitting two photometric passbands. Adopted from Strassmeier & Rice (2003). Credit: K. G. Strassmeier and J. B. Rice, A&A, 399, 315, 2003, reproduced with permission © ESO.

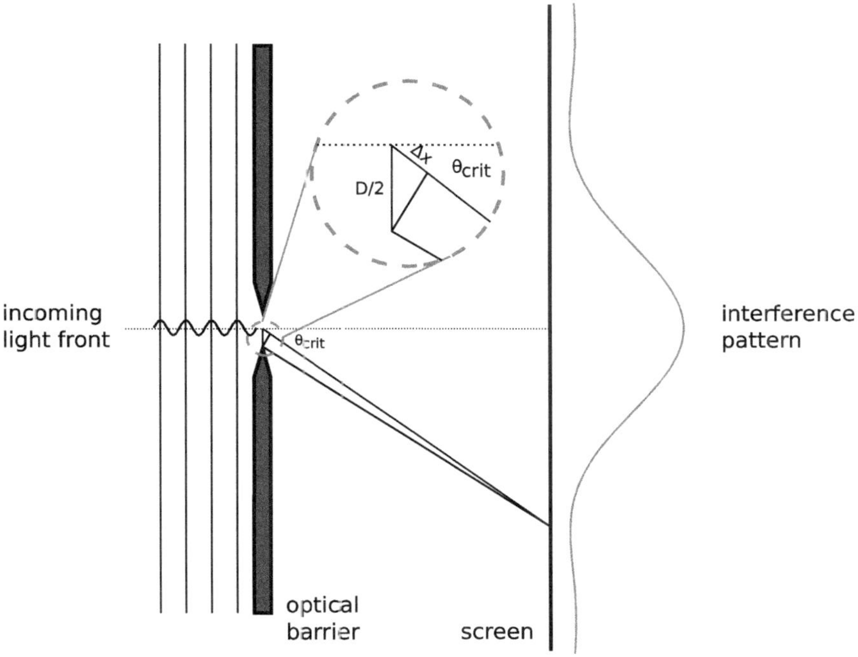

Figure 2.12. Schematic of diffraction on a finite linear aperture.

light has to travel from different parts of the aperture is different. The magnified inset in Figure 2.12 shows a critical situation where the path diffracting through the center of the aperture and the path diffracting at the edge of the aperture undergo destructive interference, resulting in null light reaching the screen. Destructive interference will occur when the light along the two paths reaches the point on screen in antiphase, i.e., where the two beams are offset in phase by π. That corresponds to the path length difference of $\Delta\lambda/2$:

$$\Delta x = n\Delta\lambda/2, \quad n \in \mathbb{N}. \tag{2.25}$$

Once we express Δx with a critical angle $\theta_{\rm crit}$ and a linear aperture width D, we arrive at the familiar diffraction criterion for the first minimum:

$$\Delta x \approx \frac{D}{2}\sin\theta_{\rm crit} \quad \Rightarrow \quad \sin\theta_{\rm crit} \approx \theta_{\rm crit} \approx \frac{\lambda}{D}. \tag{2.26}$$

The computation gets a bit trickier for 2D apertures, where we need to evaluate Fraunhofer's diffraction formula:

$$\psi(u, v) = \int\int \xi(x', y')e^{-ik(ux'-vy')}dx'dy'. \tag{2.27}$$

We used u and v for screen coordinates, x' and y' for aperture coordinates, and $k = 2\pi/\lambda$ is the wavenumber. For a circular aperture with diameter D we can rewrite the above equation in polar coordinates:

$$\psi(u, v) = \int\int \xi(r, \phi)e^{-ikr(u\cos\phi - v\sin\phi)}rdrd\phi. \tag{2.28}$$

Here $\xi(r, \phi) = 1$ for $r \leqslant D/2$ and 0 elsewhere, so if we integrate over dr from 0 to $(D/2)$ and over $d\phi$ from 0 to 2π, then $\xi(r, \phi) \equiv 1$. Finally, the term $u\cos\phi - v\sin\phi$ can be rewritten as $\beta\cos(\phi + \Delta\phi)$, where $\beta = (u^2 + v^2)^{1/2}$ and $\Delta\phi = \arctan v/u$. Thus, the integral becomes:

$$\psi(u, v) = \int_0^{2\pi}\int_0^{D/2} e^{-ikr\beta\cos(\phi+\Delta\phi)}rdrd\phi. \tag{2.29}$$

The phase shift $\Delta\phi$ is irrelevant because we are integrating over the entire ϕ range and it does not matter where we start. With a bit of mathematical weight lifting this expression integrates to:

$$\psi(u, v) \propto \frac{j_1(\rho)}{\rho} \quad \text{where} \quad \rho = k\beta D/2, \tag{2.30}$$

and where $j_1(\rho)$ is the Bessel function of the first order. This function will assume its first zero at $\rho = 1.21967\ldots \approx 1.22$. That is why the maximum resolving angle of an instrument with a circular aperture is given by the Rayleigh criterion:

$$\sin\theta_R \approx \theta_R \approx 1.22\frac{\lambda}{D}. \tag{2.31}$$

Comparing this to Equation (2.26) shows that a circular aperture differs from a linear aperture by a constant. The resolving power of a telescope with an aperture D

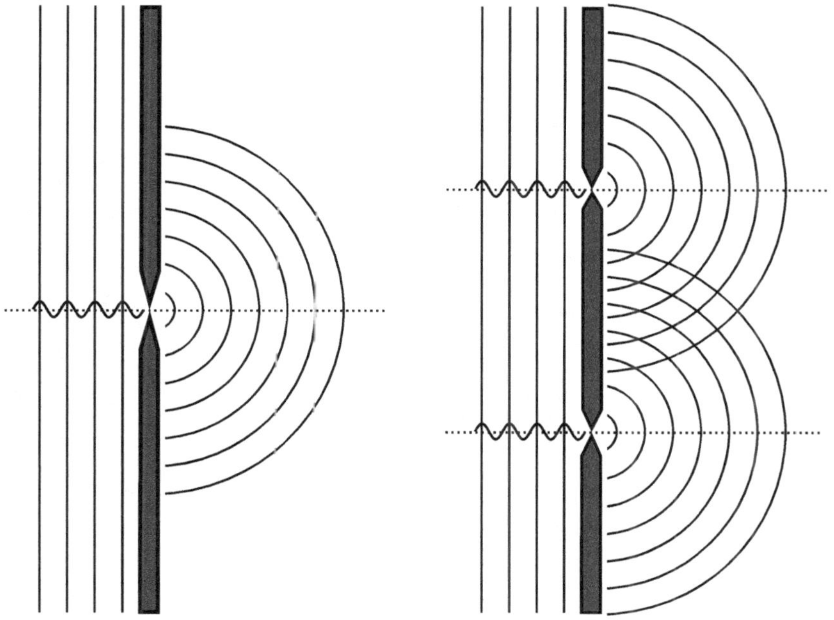

Figure 2.13. Schematic of diffraction on an infinitesimal single slit (left) and double slit (right).

in the absence of other extenuating factors (such as the atmosphere) will thus be diffraction limited as imposed by the Rayleigh criterion. For example, to resolve a 1 milliarcsecond (mas) source from space in the optical ($\lambda \sim 550$ nm), we would need a 140 m telescope. Those have proven tricky to make.

Interferometry is a way around this limitation. The idea is to combine light from an array of telescopes that are separated by a precisely known distance. The limit due to diffraction then corresponds to the size of the array rather than the size of an individual telescope aperture. For as long as the distance between the telescopes is much larger than the telescope aperture D, we can consider the telescopes as infinitesimal slits that pass a tiny amount of the overall light front coming from the source. If the finite linear aperture in Figure 2.12 were replaced with an infinitesimal aperture depicted in the left panel of Figure 2.13, then—according to the Huygens principle—the aperture itself would act as a source of concentric waves. These waves are described by a scalar[2] field:

$$u(r, t) = A(r)e^{i(kr-\omega t)}, \tag{2.32}$$

where $A(r)$ is an amplitude of the wave and ω is the wave frequency. The amplitude dependence on r can be easily deduced from energy conservation: as the wave propagates outward, it needs to carry the same net amount of energy per unit time

[2] Electromagnetic radiation is, of course, a vector field, i.e., a solution to the wave equation $(c^2\nabla^2 - \partial^2/\partial t^2)\,\boldsymbol{u} = 0$, but since concentric waves only have a radial dependence, the wave equation that describes it, $(c^2\partial^2/\partial r^2 - \partial^2/\partial t^2)\,u = 0$, is solved by a scalar field.

(i.e., constant power). The intensity of the wave is *by definition* given as the amount of energy per unit time per unit area, i.e., it is proportional to the square of the field: $I \propto uu^* \equiv A^2(r)$. Thus,

$$P = I\left(\frac{1}{2}4\pi r^2\right) \propto A^2(r)2\pi r^2 \quad \Rightarrow \quad A(r) \propto \sqrt{\frac{P}{2\pi}}\frac{1}{r}. \tag{2.33}$$

Light radiation coming from an infinitesimally small aperture is thus:

$$u(r,\, t) \propto \frac{1}{r}e^{i(kr-\omega t)}. \tag{2.34}$$

If we have two slits, depicted in the right panel of Figure 2.13, the concentric wavefronts will undergo interference. Each slit acts as a source of its own concentric waves, and the combined field at any point is obtained by superposition:

$$u(r,\, t) = u(r_1,\, t) + u(r_2,\, t) \propto \frac{1}{r_1}e^{i(kr_1-\omega t)} + \frac{1}{r_2}e^{i(kr_2-\omega t)}, \tag{2.35}$$

where r_1 and r_2 are spatial distances from the first slit and from the second slit to the point of interest, respectively. If we denote the difference in path length as $\Delta r = r_2 - r_1$ and at the same time assume that r_1 and r_2 are much larger than the separation between the slits so that $r_1/(r_1 + \Delta r) \approx 1$, the combined field can be written as:

$$\begin{aligned} u(r,\, t) &\propto \frac{1}{r_1}e^{ikr_1}e^{-i\omega t}\left(1 + \frac{r_1}{r_1 + \Delta r}e^{ik\Delta r}\right) \\ &\approx \frac{1}{r_1}e^{ikr_1}e^{-i\omega t}(1 + e^{ik\Delta r}). \end{aligned} \tag{2.36}$$

The intensity of the combined field is then:

$$I \propto uu^* = \frac{1}{r_1^2}(1 + e^{ik\Delta r})(1 + e^{-ik\Delta r}) = \frac{2}{r_1^2}(1 + \cos k\Delta r). \tag{2.37}$$

Following the same argument as in Figure 2.12, the condition for constructive interference ($\cos k\Delta r = 1,\ k\Delta r = 2\pi$) then immediately leads to the on-screen angular fringe separation, $\theta = \lambda/B$, where B is the separation between the slits. The resolution of the two-slit interferometer is limited by the condition for destructive interference ($\cos k\Delta r = -1$, $k\Delta r = \pi$), i.e., $\theta_{\min} = \lambda/2B$.

If the incident wavefront is displaced by an angle ϕ, the fringe pattern will be displaced by the same angle ϕ. To see this, note that the only circumstance that changes with respect to perpendicular incident wavefront is the phase at which light hits the two slits. While in the case of perpendicular incidence the phase of the wave was the same at both slits, that is (generally) no longer true for inclined wavefronts. The linear offset along the wave, $\Delta r' = B\phi$, is added to the path length difference Δr to form the complete phase shift between the two slits.

This is important when considering extended emitters: the combined wavefronts from point sources that compose an extended emitter smear the fringes by the amount commensurate with the size of the emitter. In the worst possible case, when the angular size of the emitter equals λ/B, the fringes will "wash out" completely, i.e., the interference will result in a constant angular illumination. To quantify this, we introduce interferometric visibility:

$$V = \frac{\text{fringe amplitude}}{\text{average intensity}} \equiv \frac{I_{\max} - I_{\min}}{I_{\max} + I_{\min}}, \tag{2.38}$$

where $I_{\min}$ and $I_{\max}$ correspond to the minimum and maximum fringe intensity, respectively. When the angular size of the source is much smaller than λ/B, $V \to 1$; as the size approaches λ/B, $V \to 0$. Figure 2.14 recaps these key points; the corresponding animations can be viewed in the interactive version of the book or downloaded from the book homepage.

We, of course, cannot control the angular size of the source we are observing, but we can control the separation between the slits to affect the interferometric visibility. The "slits" in astronomical applications are individual telescopes, and the distance between them, called the *baseline* B, is a quantity that we control. The mapping between intensity $I(\theta)$ and interferometric visibility $V(B)$ is a simple Fourier transform. This is known as the Van Cittert–Zernicke theorem (Thompson et al. 2001).

Combining light from individual telescopes is not without its challenges. Interferometric data are not images by themselves; instead, the images need to be reconstructed from spatially and temporally combined fringe patterns recorded by interferometers. This constitutes a signal processing problem that is solved by parametric forward-modeling, which we describe next.

Consider two telescopes pointing at the same source in the direction $\boldsymbol{S}$, separated by a baseline $\boldsymbol{B}$ (see Figure 2.15). The field at each of the telescopes is:

$$u_1(r, t) = Ae^{ik(D+\boldsymbol{B}\cdot\boldsymbol{S})}, \qquad u_2(r, t) = Ae^{ik(D+\delta)}, \tag{2.39}$$

where A is the (complex) amplitude of the signals and δ is an arbitrary delay that is purposefully introduced to one of the signals (in this case u_2) at the instrument's combiner in order to control the interference. You may also notice that we omitted the temporal term $-\omega t$: as intensity is defined as a temporal average, that term does not play a role. The intensity measured at the combiner is given by the time-averaged square of the combined field amplitude:

$$\begin{aligned} I &= \langle (u_1 + u_2)(u_1 + u_2)^* \rangle \\ &= 2AA^* \left\langle \left[1 + \frac{1}{2}(e^{ik(\boldsymbol{B}\cdot\boldsymbol{S}-\delta)} + e^{-ik(\boldsymbol{B}\cdot\boldsymbol{S}-\delta)}) \right] \right\rangle \\ &= 2I_0 \, \langle [1 + \cos(k(\boldsymbol{B}\cdot\boldsymbol{S} - \delta))] \rangle, \quad \text{where} \quad I_0 \equiv AA^*. \end{aligned} \tag{2.40}$$

The term $\Delta \equiv \boldsymbol{B}\cdot\boldsymbol{S} - \delta$ is the *optical path difference*. If $\Delta = 0$, interference is constructive; if $\Delta = \lambda/2$, interference is destructive.

Figure 2.14. Interference patterns and time-averaged intensity fringes. The wavefront is arriving from infinity on the left, hitting a barrier depicted in dark red and being transmitted through the slits toward the screen on the right. In all figures the $1/r$ intensity attenuation is not depicted; only the interference fringes are depicted. Top: a single infinitesimal slit (left) and two infinitesimal slits (right). A single slit will produce no fringes. A double slit does produce the fringes that would be equally spaced if the screen were curved with a radius R; the fringe separation depicted here is a consequence of projecting onto a flat screen. Middle: the effect of the incident wavefront angle for $-5°$ (left) and for $+5°$ (right). The fringes are displaced by the same angle and are close to the antiphase. Bottom: the effect of an extended source size on interferometric visibility. The size of the source is depicted by a dark red bar on the left-hand side. The visibility is either reduced (left) or diminished (right). Animations of the images are available online at https://doi.org/10.1088/978-0-7503-1287-5. The animations can also be downloaded as stand-alone files from the book homepage.

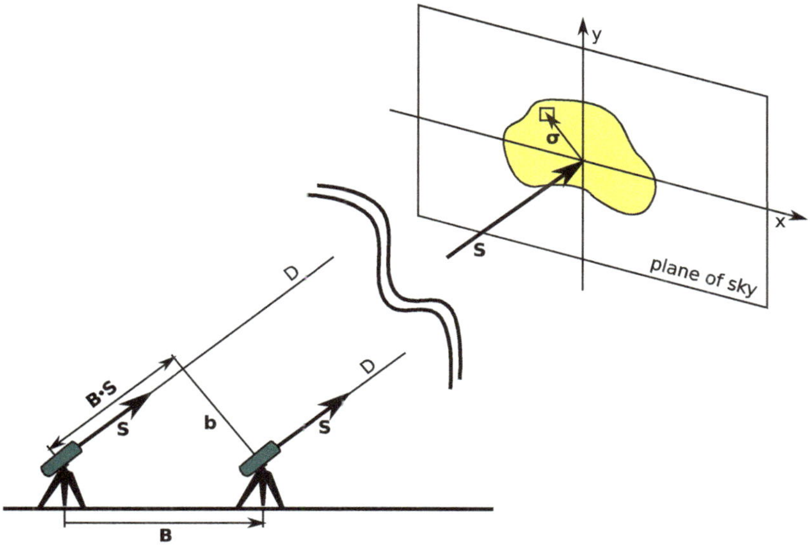

Figure 2.15. Two telescopes observing an extended light source. The delay in contemporary light pulse emitted from the source is due to the added distance $\boldsymbol{B} \cdot \boldsymbol{S}$, causing interference.

As the source in the sky is extended, in addition to the optical path difference we need to take into account the path length change because of the spatial distance $\boldsymbol{\sigma}$ in the plane of sky (see Figure 2.15). The electromagnetic field at the telescopes is now:

$$u_1 = Ae^{ik[D-\boldsymbol{B}\cdot(\boldsymbol{S}+\boldsymbol{\sigma})]}, \qquad u_2 = Ae^{ik(D+\delta)}, \tag{2.41}$$

and the intensity becomes:

$$I = 2I_0 \left\langle 1 + \cos(k(\Delta + \boldsymbol{B} \cdot \boldsymbol{\sigma})) \right\rangle. \tag{2.42}$$

The problem? The atmosphere. The incoming wavefront is dispersed by the turbulence in the atmosphere, causing the image resolution to be limited by the coherence scale of the wavefront instead of the size of the telescope. For example, the transverse distance $r_0(\lambda)$ at which coherence is substantially deteriorated is $\sim$10 cm at $\lambda = 500$ nm (Monnier 2003). In other words, as soon as the size of the telescope exceeds $\sim$10 cm, the interferometric image will be limited by seeing rather than by diffraction. The observed angular size of a point source will thus be driven almost exclusively by seeing. To make things worse, the atmosphere is not static, and turbulent cells change with time on a typical timescale $t_0(\lambda)$ called the *coherence timescale*. At $\lambda = 500$ nm the typical coherence timescale is $\sim$10 ms (Monnier 2003). The two parameters, $r_0(\lambda)$ and $t_0(\lambda)$, limit the useful light-gathering size and exposure time of an interferometer, and they depend greatly on the observing site. The turbulence basically breaks the Van Cittert–Zernicke theorem, and we are left with model-dependent solutions to reconstruct the image.

The most commonly used strategy is the introduction of the controlled phase delay δ in Equation (2.39). The idea is that the sum of phases over a closed baseline loop is independent of the atmosphere-induced phase shifts. To see this, consider a three-telescope interferometer. The phase difference between each telescope depends on the intrinsic delay and on atmospheric delay:

$$\begin{aligned}\Delta\Phi_{obs}(1-2) &= \Delta\Phi_{intrinsic}(1-2) + \Delta\Phi_{atm}(2) - \Delta\Phi_{atm}(1),\\ \Delta\Phi_{obs}(2-3) &= \Delta\Phi_{intrinsic}(2-3) + \Delta\Phi_{atm}(3) - \Delta\Phi_{atm}(2),\\ \Delta\Phi_{obs}(3-1) &= \Delta\Phi_{intrinsic}(3-1) + \Delta\Phi_{atm}(1) - \Delta\Phi_{atm}(3).\end{aligned} \tag{2.43}$$

The sum of the three terms cancels the atmospheric terms:

$$\Delta\Phi = \Delta\Phi_{intrinsic}(1-2) + \Delta\Phi_{intrinsic}(2-3) + \Delta\Phi_{intrinsic}(3-1). \tag{2.44}$$

Thus, the sum of the three fringe phases along the baseline loop is a good interferometric observable. This strategy is called the *closure phase* method (Readhead et al. 1988). Image reconstruction then boils down to finding an image that fits interferometric visibility amplitudes and closure phases. This is an inversion problem, so there are infinite images that will produce the same visibilities and closure phases, and additional constraints need to be imposed to select the "likeliest" image. These constraints are sophisticated (Thiebaut & Giovannelli 2010; Baron 2016), and they are built into image reconstruction software such as CLEAN (Högbom 1974) and MEM (Pina & Puetter 1992). In general, though, the more telescopes we have within the interferometer, the better the recovery of intrinsic phase distributions. For 21 telescopes, ~90% of the phase information can be reconstructed (Monnier 2003). The currently available optical interferometric imaging facilities include the Center for High-Angular Resolution Astronomy (CHARA; ten Brummelaar et al. 2005), the Very Large Telescope Interferometer (VLTI; Haguenauer et al. 2008), the Magdalena Ridge Observatory Interferometer (MROI; Creech-Eakman et al. 2008), and the Navy Prototype Optical Interferometer (NPOI; Benson et al. 2004). Figure 2.16 depicts CHARA's *H*-band reconstructed image of ε Aurigae, an F-type supergiant that eclipsed a large dark disk every ~27 yr (Kloppenborg et al. 2010).

2.5 Astrometry

The basic task of astrometry is to provide positions, space motions, and trigonometric distances to observed objects with respect to a fixed coordinate system. Its beginnings trace back to ancient Greece, China, and the Middle East, when seafarers and early astronomers measured the positions of stars and planets in the sky using astrolabes and armillary spheres (see Figure 2.17). The basic idea was to measure angles (typically altitude and azimuth) of stars and use those angles for navigation. In modern times, astrometry does much more: it establishes a celestial reference frame for describing the motion of objects in the sky, and it provides distances to the object by measuring the parallax as Earth travels around the Sun. In terms of binary and multiple stars, we can observe orbital motion of resolved components as a function of time, thus obtaining the absolute scale of the system.

As astrometry relies on resolving power of instrumentation, it is diffraction limited, although for the ground-based instruments the limiting factor in most cases is atmospheric seeing. That is why, until recently, astrometry was not particularly

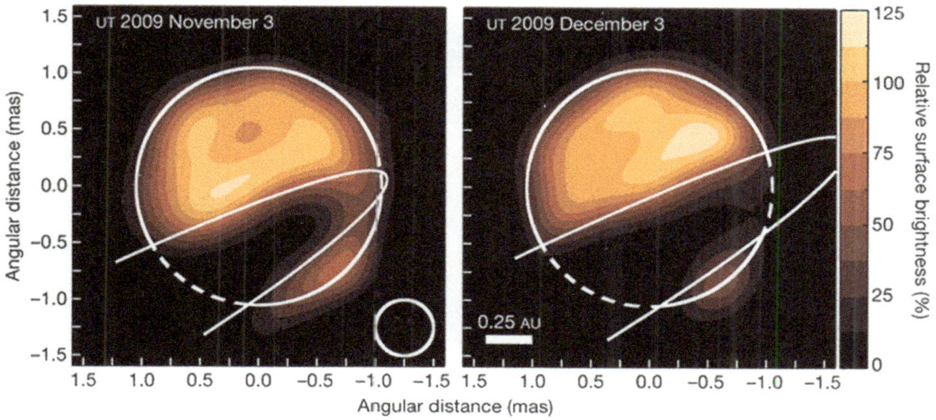

Figure 2.16. Reconstructed image of ε Aurigae from CHARA *H*-band observations. The model is overplotted in white. The circle in the bottom right of the November 3 image corresponds to the CHARA *H*-band resolution of 0.5 mas. Adopted from Kloppenborg et al. (2010). Reprinted by permission from Nature: Springer Nature, International Journal of Science, "Infrared Images of the Transiting Disk in the ε Aurigae System," B. Kloppenborg et al., 2010.

Figure 2.17. Left: monument to Nicolaus Copernicus in Torun, Poland. Copernicus is holding a compass and an armillary sphere. Photo courtesy of Milena Ratajczak. Right: medieval astrolabe, ca. 1345–1355, from the Arab World Institute in Paris. Photo courtesy of Frank Maloney.

well suited for objects that are not in our immediate vicinity. However, with the advent of interferometry in the 1970s, and once the astrometric instruments were lifted above the atmosphere, a major limitation to resolving distant objects is removed. Pioneering results of the *Hipparcos* satellite (Perryman et al. 1997) have reached the accuracy of 1 mas, on par with the best ground-based optical interferometric accuracy (Harris et al. 2016). The ongoing space mission *Gaia* (Brown et al. 2016a) is equipped with an astrometric instrument that is achieving

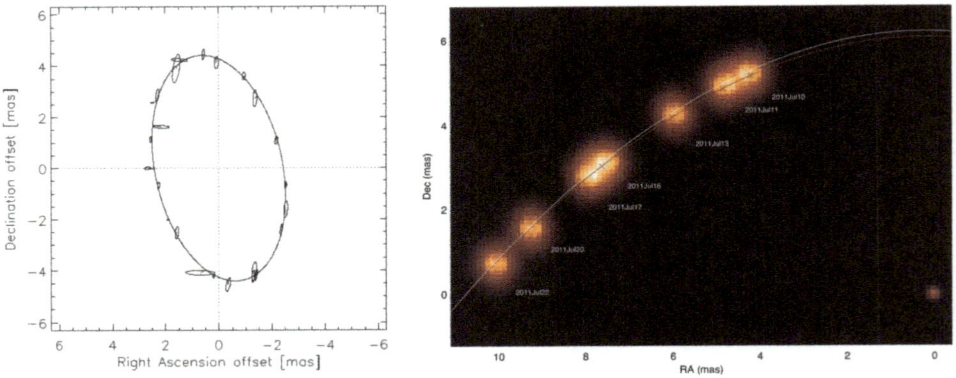

Figure 2.18. Left: astrometric orbit of o Leonis, determined from interferometric data acquired with Mark III and NPOI instruments. Reproduced from Hummel et al., "Orbital and Stellar Parameters of Omicron Leonis from Spectroscopy and Interferometry," The Astronomical Journal, 121.3, 623-1635, 2001. © 2001. The American Astronomical Society. Right: interferometric observations of δ Scorpii during periastron fly-by, observed by CHARA. Reproduced from Che et al., "Imaging Disk Distortion of Be Binary System δ Scorpii near Periastron," The Astrophysical Journal, 757.1, 29, 2012. © 2012. The American Astronomical Society.

microarcsecond (μas) accuracy, which is unprecedented and is bound to place astrometry in the prominent place of modern data acquisition methods.

At the ~arcsecond level, relativistic effects discussed in Section 2.1 are already discernible, and at the μas level they are absolutely crucial. That is why space astrometry uses the BCRS to describe the motions of observed objects. The corrections built into the BCRS suffice for μas astrometry, but they will no longer suffice for nanoarcseconds, where additional phenomena such as optical interstellar scintillation, geometric cosmology, and gravitational wave spacetime ripples become important (van Altena 2013).

Astrometric proper motions of visual binaries have long been utilized with remarkable success for obtaining accurate masses of individual stars (Jankov et al. 2014). Apart from visual binaries, astrometric data, when accompanied by the RV observations, can also provide masses from the photocenter motions of double-lined spectroscopic binaries purely from observations, without any underlying physical model. Figure 2.18 shows two examples: o Leonis, an F9III–A5V double star, where RV and interferometric observations in conjunction provided masses of the companions to ~0.5% and the distance to ~0.3% (Hummel et al. 2001), and δ Scorpii, a highly eccentric ($e = 0.94$) system observed close to periastron passage and leading to a sub-1% accuracy in the inclination and ~500 ppm in semimajor axis and eccentricity of the system (Che et al. 2012).

2.6 Polarimetry

The polarization of light refers to a preferred direction of the electric component of the field as light propagates through space. It is inherent to the source of light or the properties of the medium through which the light propagates. It suffices to describe polarization with only the electric component of the field because the magnetic component follows uniquely from Maxwell's equations.

Light can be unpolarized, partially polarized, or fully polarized. The polarization is parameterized by the Stokes parameters (Stokes 1852; Tinbergen 2005) that determine the amount of radiant energy per unit area per unit wavelength per unit solid angle, *averaged over time*:

$$S = \begin{pmatrix} I \\ Q \\ U \\ V \end{pmatrix} = \begin{pmatrix} A^2 \\ A^2 \cos 2\beta \cos 2\xi \\ A^2 \cos 2\beta \sin 2\xi \\ A^2 \sin 2\beta \end{pmatrix} \equiv I \begin{pmatrix} 1 \\ \cos 2\beta \cos 2\xi \\ \cos 2\beta \sin 2\xi \\ \sin 2\beta \end{pmatrix}, \tag{2.45}$$

where A is the amplitude of the electric field, and angles β and ξ are the angles with respect to the reference frame (Figure 2.19). The factor 2 in 2β and 2ξ is there by convention, to illustrate that a rotation of the ellipse by 180° about the center and a rotation of the ellipse by 90° and transposition of the semimajor and semiminor axes will both result in an indistinguishable ellipse.

The Stokes parameters are closely related to fundamental quantities. Parameter I is the intensity, while the parameters Q, U, and V are spherical coordinates that determine the type of polarization. For example, a 100% linear polarization is described by the Stokes vector $S_{\text{lin}} = A^2(1, \cos 2\xi, \sin 2\xi, 0)$, where ξ is the polarization angle. A 100% circular polarization is described by the Stokes vector $S_{\text{circ}} = A^2(1, 0, 0, \pm 1)$, where the sign of V determines the chirality of polarization. A 100% elliptical polarization has all four Stokes parameters nonzero, but they are not all independent; instead, they are constrained by $Q^2 + U^2 + V^2 = I^2$, which is clearly evident from Equation (2.45). If light is unpolarized, the time averages over Q, U, and V are all 0, so the Stokes vector is $S_{\text{unp}} = A^2(1, 0, 0, 0)$.

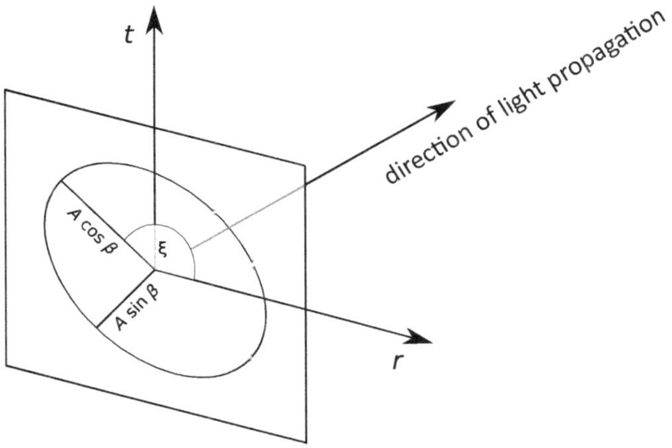

Figure 2.19. Parameterization of polarization by angles ξ and β. ξ is the polarization angle, and $\tan \beta$ is the axial ratio of the ellipse. Directions r and t are perpendicular to the direction of light propagation s and are selected by practical considerations, i.e., alt-az, equatorial, galactic, etc.

A partially polarized light can be written as a combination of a 100% polarized light and unpolarized light:

$$I\begin{pmatrix}1\\Q\\U\\V\end{pmatrix} = I\begin{pmatrix}p\\Q\\U\\V\end{pmatrix} + I\begin{pmatrix}1-p\\Q\\U\\V\end{pmatrix}, \tag{2.46}$$

where $p = \sqrt{Q^2 + U^2 + V^2}/I$ is the *degree of polarization*. If $p = 1$, the light is 100% polarized; if $p = 0$, the light is unpolarized; for $0 < p < 1$, light is partially polarized.

As we have established, the Stokes parameters describe the flow of radiant energy. When light propagates through the medium, the medium itself can change the polarization of light. Mathematically we describe that by the Müller matrices (Müller 1948):

$$\boldsymbol{S}' = \begin{pmatrix} m_{11} & m_{12} & m_{13} & m_{14}\\ m_{21} & m_{22} & m_{23} & m_{24}\\ m_{31} & m_{32} & m_{33} & m_{34}\\ m_{41} & m_{42} & m_{43} & m_{44}\end{pmatrix}\boldsymbol{S} \equiv \boldsymbol{M}\cdot\boldsymbol{S}. \tag{2.47}$$

As the radiation travels through several media, we simply need to multiply the corresponding Müller matrices of each medium, keeping in mind that the matrices do not commute, i.e., the last matrix corresponds to the first traversed medium. Müller matrices are also very practical to describe instrumental response; for example, if a linear polarization filter (called the analyzer) is inserted into the light beam, its effect on total light can be readily described by:

$$\boldsymbol{M}_{\rm lin}(\xi = 0) = \begin{pmatrix} 1 & 1 & 0 & 0\\ 1 & 1 & 0 & 0\\ 0 & 0 & 0 & 0\\ 0 & 0 & 0 & 0\end{pmatrix}. \tag{2.48}$$

This can be quickly verified if we multiply $\boldsymbol{S}_{\rm unp} = I(1, 0, 0, 0)$ by $\boldsymbol{M}_{\rm lin}(\xi = 0)$ to obtain $\boldsymbol{S}_{\rm lin} = I(1, 1, 0, 0)$.

You may have noticed that there is no notion of phase in the Stokes parameters: so far we only used the time-averaged quantities. If phase were ever of interest, for example, when combining polarimetry with interferometry, an alternative formalism needs to be utilized. The formalism, developed by Jones (1941), is based on the complex notation of the fields (much like what we discussed in Section 2.4). The discussion of the Jones formalism is beyond the scope of this brief review; interested readers should consult Tinbergen (2005) and other books that focus on polarimetry alone.

In the astronomical context, the degree of polarization coming from celestial sources is often low, up to a few percent (Kolokolova et al. 2015). Most light is unpolarized: each photon, of course, has a particular polarization, but if an ensemble of photons has random polarizations, the net polarization is 0. In general, polarization arises from asymmetry and anisotropy of the observed astronomical source or of the medium that the light traverses. The typical sources of polarization in close binaries are emergent light scattered on the star's photosphere due to shape distortion; light of one star scattered at the surface of the other star (reflection effect); light scattered by the accretion disk, dust, streams, and/or circumstellar clouds; electron scattering in gas flows, flares, and high-temperature atmospheres; magnetic surface fields; dichroic interstellar extinction; thermal IR emission from the aligned elongated dust grains; line scattering, etc. Some of these and others are discussed by Piirola (2010) and Bjorkman (2012).

A basic polarimetric measurement is equivalent to photometry, with an additional polarimetric analyzer inserted in the input beam. This analyzer transmits only a fraction of the light with the matching polarization. By rotating the polarimetric analyzer, the degree of polarization may be measured for the given source. To avoid systematic bias and drift errors, polarimetric observations are usually provided in *fractional* polarization:

$$\mathcal{P} = \frac{F_1 - F_2}{F_1 + F_2}, \tag{2.49}$$

where F_1 and F_2 are observed fluxes at two orthogonal positions of the polarimetric analyzer. Typically one finds the minimal and maximal values of the transmitted flux:

$$\mathcal{P}_{\max} = \frac{F_{\max} - F_{\min}}{F_{\max} + F_{\min}}. \tag{2.50}$$

Other polarimetric measurements include spectropolarimetry, i.e., the measurement of polarized SEDs, and circular polarimetry, specialized for the study of magnetic fields.

Since polarimetric observations discard significant portions of the input flux, considerable observing time on a large telescope is generally needed in order to reduce the photometric noise. Polarimetry should therefore not be used indiscriminately, but only when it provides insights that other methods cannot give.

As an example of polarimetry in action, Berdyugin et al. (2016) investigated polarization in the O-type binary system HD 48099. They found a clear polarization signal (Figure 2.20) with a period that is twice the orbital period, typical of binary stars. They successfully reproduced the data by three models: reflection off the photospheres, a colliding wind model, and an accretion stream model. These are fundamentally different models, and additional data are required to discriminate between them; the authors rejected the accretion stream model based on the lack of any other evidence for mass transfer and proposed that scattering and wind collision are the most likely scenarios responsible for the polarimetric signature. Thus, polarimetry certainly has the potential to contribute to our understanding of close binary and multiple stars, mostly in conjunction with other observational methods.

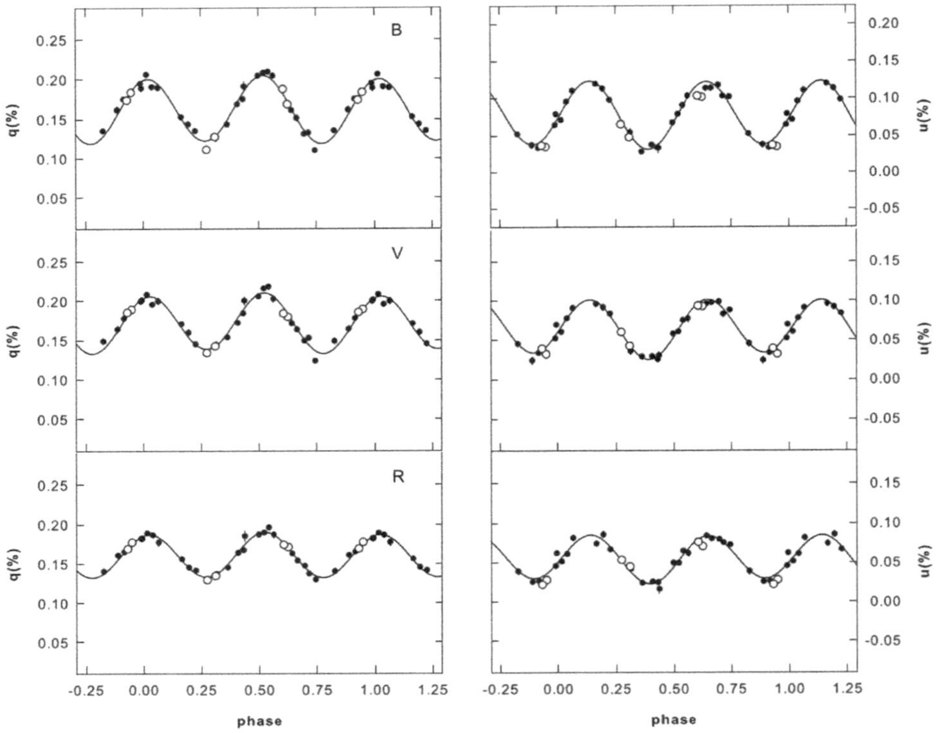

Figure 2.20. Polarimetric observations of HD 48099 as a function of orbital phase. Stokes parameters Q (left) and U (right) are given for three photometric passbands. Symbols correspond to two different instruments. The error bars are smaller than the symbol sizes. Credit: Berdyugin et al., A&A, 591, A92, 2016, reproduced with permission © ESO.

2.7 Asteroseismology

Asteroseismology is the study of internal stellar structure by way of seismic wave propagation observed in power spectra. This wave propagation—stellar pulsations—arises from physical instabilities inside the star (Aerts et al. 2010). Pulsations reveal themselves through photometric and/or spectroscopic variations, and depending on the number of pulsational modes, their interpretation can be arbitrarily complex. For the Sun we have the luxury of sampling pulsation components *spatially*, across the disk, but for other stars we observe integrated light instead. In practice, that means that, due to cancellation effects, we are limited to observing low-order modes in unresolved stars.

Pulsations in spherical stars can be described by a set of three wavenumbers: radial order n, nonradial degree l, and longitudinal order m. Radial order n gives the number of nodes between the center and the surface of the star, nonradial degree l gives the number of all (longitudinal and latitudinal) node lines, and longitudinal order m gives the number of longitudinal node lines (Aerts et al. 2010). Longitudinal order m can assume values from $-l$ to l, where $\pm m$ modes correspond to mode chirality and are indistinguishable for spherically symmetric stars because there is no preferred axis, but different for nonsymmetric (i.e., rotating, distorted) stars where axial asymmetry breaks the degeneracy. Thus, $l = 0$ modes are radial pulsations, $l = 1$ modes are dipole modes, $l = 2$ are quadrupole modes, etc. Figure 2.21 depicts the first six surface modes (up to $l = 2$) with grossly exaggerated amplitudes for clearer visualization. These plots are animated in the interactive version of this book.

Stars pulsate for many different reasons. The pulsations in δ Sct stars, Cepheids, and RR Lyrae stars are driven by the κ mechanism (Saio 1993), where stellar interior opacity for radiative transfer of heat is a strong function of temperature. As the star expands, the blocking layer inside the star cools and becomes more opaque; energy is trapped, and the star contracts. As it contracts, the blocking layer heats up and becomes less opaque, enabling energy to escape. Solar-like oscillators are stochastic, driven by surface convection (Christensen-Dalsgaard & Daeppen 1992), where the turbulent flow of fluids excites and dampens oscillations. γ Dor pulsations are caused by convective blocking (Guzik et al. 2000), where the inner edge of the convection zone is sharp and the convection timescale is longer than the pulsation timescale, so perturbations grow into large, coherent pulsations. Heartbeat stars—highly

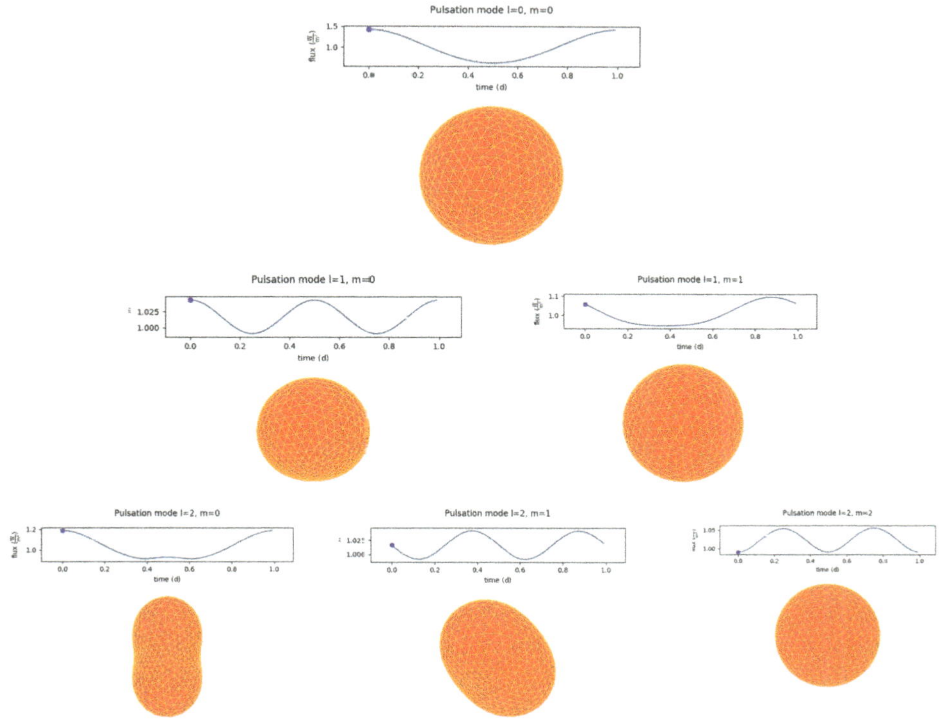

Figure 2.21. First few pulsational modes described by the nonradial degree l and longitudinal order m. The top panels depict flux variation as a function of time. The image at the bottom shows the shape of the star at maximum distortion amplitude. The interactive version of these plots shows the animation of pulsation modes and the corresponding light curves. The animation is available online at https://doi.org/10.1088/978-0-7503-1287-5.

eccentric binary stars—frequently feature tidally induced pulsations (Kumar et al. 1995; Thompson et al. 2012). At periastron fly-by, gravitational interaction "pinches" the stars for the duration that excites one of the otherwise unexcited eigenmodes by way of resonance.

Depending on the restoring force, pulsations in stars are either pressure (p) modes, where the restoring force is pressure gradient, or gravity (g) modes, where the restoring force is buoyancy. Pressure modes are caused by acoustic waves that propagate through gas by compression and decompression, while gravity modes are caused by density waves that propagate through gas by adiabatic expansion. The typical frequencies of p- and g-modes depend on the size of the respective "cavity" inside the star; for the Sun, p-modes are between $\sim$170 and 350 cycles day^{-1} (2–4 mHz; also known as the 5-minute oscillations; García et al. 2001), while g-modes are between 0 and $\sim$35 cycles day^{-1} (0–0.4 mHz; Appourchaux et al. 2010). Since convection dampens g-modes by making them evanescent, we do not expect to see them on the surface of the Sun. The p-modes, on the other hand, do propagate through convective envelopes and show up on the surface for as long as their frequency is smaller than the critical frequency $\nu_{\rm ac}$. This frequency, called the acoustic cutoff frequency, delimits waves that are trapped inside the cavity and those that propagate through the atmosphere. The acoustic cutoff frequency for the Sun is $\sim$458 cycles day^{-1} (5.3 mHz; Jiménez et al. 2011). For hotter stars with thin convective and radiative envelopes we expect to see both p- and g-modes at the surface.

Stochastic pulsation modes are frequently observed to have a high radial order (n) wavenumber and a low nonradial degree (l) wavenumber (Ledoux et al. 1961). In the Sun, for example, the highest-amplitude modes have n in the range of 19–22 and l up to 4 (Bedding 2014). This is important because, in the $n \gg l$ limit, we can largely neglect the perturbation of the gravitational potential. Within this approximation, first proposed by Cowling (1941), p-modes are expected to be equally spaced in frequency and g-modes are expected to be equally spaced in period (Tassoul 1980). Figure 2.22 shows the observed power spectra of the Sun and several stars that exhibit solar-like pulsations. Note that the term "solar-like" pertains to *pulsations*, not the types of stars. In particular, ξ Hya is a giant, and η Boo and β Hyi are subgiants, yet they still exhibit solar-like (stochastic, convection-driven) pulsations. Power spectra are plotted in log-frequency (left) to highlight the dependence of peak amplitudes on the evolutionary state and in $\nu/\nu_{\rm ac}$ (right) to demonstrate the dependence of mode spacing on evolutionary state.

A closer look at a power spectrum of solar-like p-mode pulsations reveals regular patterns (see Figure 2.23). Modes with equal nonradial degree l and radial orders n that differ by 1 (i.e., modes (20, 0) and (21, 0), or (21, 1) and (22, 1), etc.) are separated by a fixed amount that we call *large frequency separation* and denote with $\Delta\nu$. Modes with different nonradial degree l and radial orders n that differ by 1 (i.e., modes (20, 0) and (21, 1), or (21, 0) and (22, 2), etc.) are called *small frequency separations*, and we denote them with $\delta\nu_{ll'}$. The identification of the modes is done by comparing the observed power spectrum to theoretical predictions from pulsation codes such as Gyre (Townsend et al. 2014).

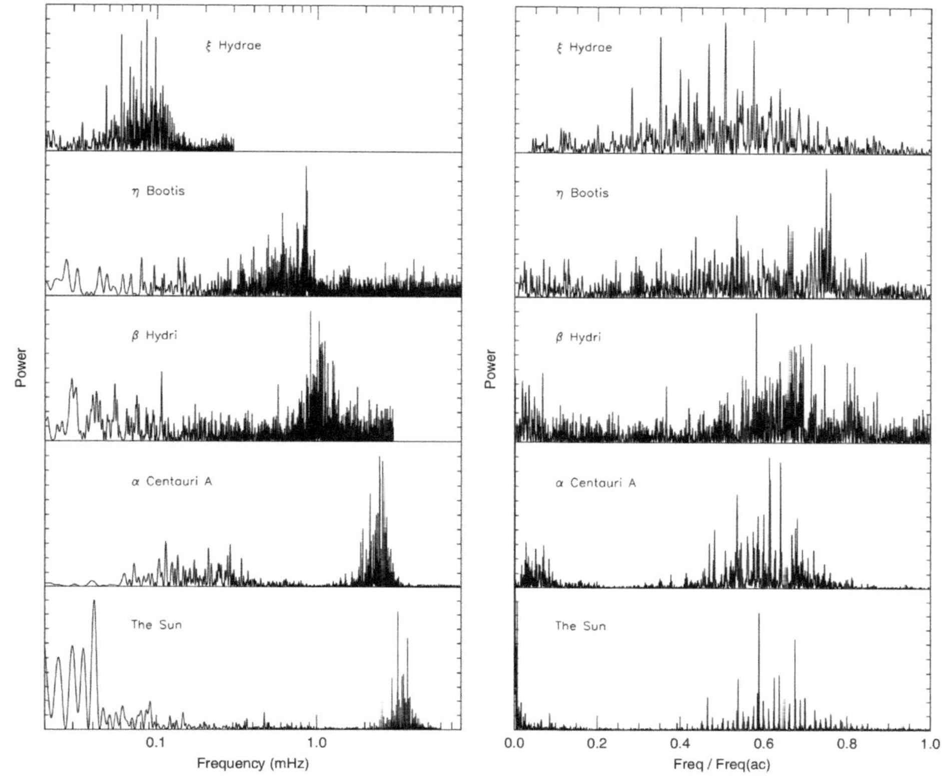

Figure 2.22. Observed power spectra of the Sun and four stars that feature solar-like pulsations, ranging from giants (top) to main-sequence stars (bottom). On the left, the horizontal axis is observed frequency on a logarithmic scale. On the right, the frequency is plotted on a linear scale as a fraction of the acoustic cutoff frequency. The y-axis scale has been made uniform to facilitate comparison; in reality, the amplitude increases from bottom to top. Adopted with permission from Bedding, T., and Kjeldsen, H., "Solar-like Oscillations," 20.2, 203-212, Publications of the Astronomical Society of Australia, © 2013 Cambridge University Press.

These observables—$\nu_{\max}$, $\Delta\nu$, and $\delta\nu_{ll'}$—can be directly related to the physical quantities. The large frequency separation $\Delta\nu$ corresponds to the inverse of the sound-travel time from the surface of the star to the core and back (Christensen-Dalsgaard 2003):

$$\Delta\nu = \left(2\int_0^R \frac{dr}{c}\right)^{-1}, \tag{2.51}$$

where $c = \sqrt{\Gamma_1 p/\rho}$ is the speed of sound, Γ_1 is the adiabatic exponent, p is pressure, and ρ is density. Following the ideal gas law, ρ is proportional to $p\mu/T$, where μ is the molar mass and T is the temperature. It follows then that $c \propto \sqrt{T/\mu}$, and since the central temperature can be estimated as $T \propto \mu M/R$ (Kippenhahn & Weigert 1994), we conclude that $\Delta\nu \propto (M/R^3)^{1/2}$. The large frequency separation is thus a

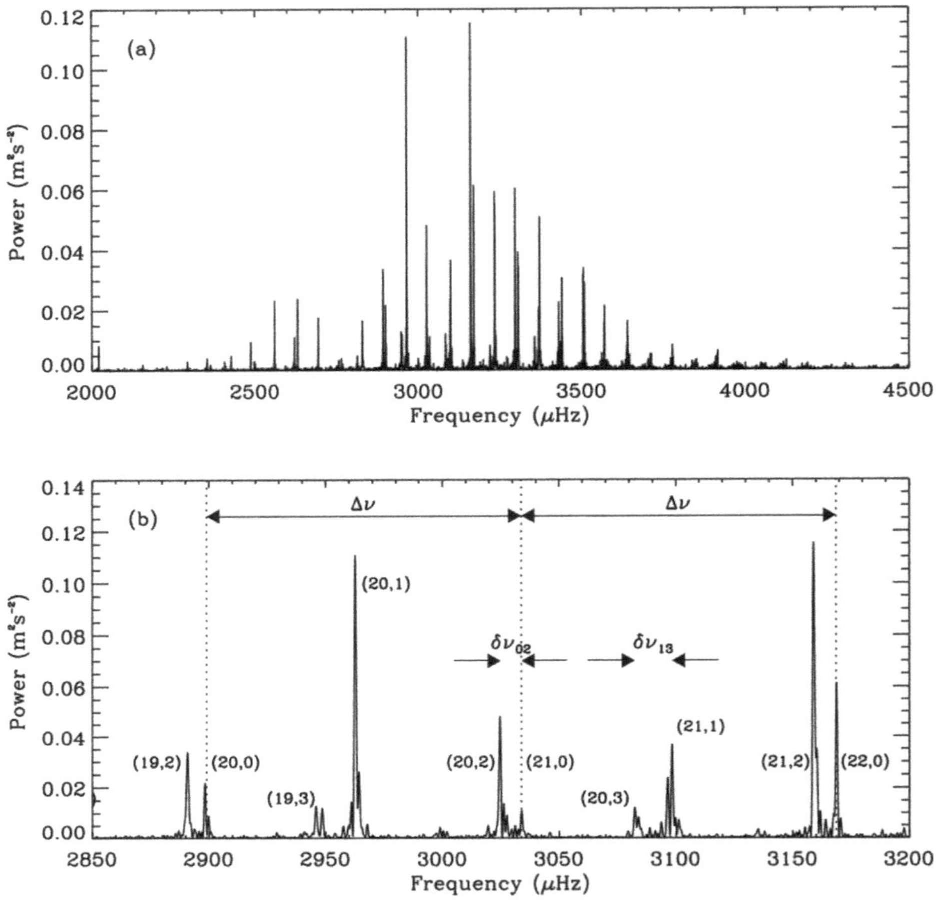

Figure 2.23. Solar power spectrum from 10 days of disk-integrated velocity measurements with BiSON (Broomhall et al. 2009). Top: the "comb" of pulsations that peaks at $\nu_{\max} \sim 3100\,\mu$ Hz. Bottom: a magnified portion of the power spectrum with large and small frequency separations marked. Adopted from Bedding (2014). Permission taken from Bedding, T., "Solar-like Oscillations: An Observational Perspective," pp. 60–86, in P. Pallé & C. Estenan (eds.), Asteroselsmology © 2014 Cambridge University Press.

direct measure of the mean stellar density. The small separations are sensitive to the sound speed gradient near the core, and hence to the age of the star (Bedding 2014). The frequency of the peak with maximum power $\nu_{\max}$ is related to the driving mechanism. Brown (1991) worked out that $\nu_{\max}$ is proportional to the acoustic cutoff frequency $\nu_{ac} = cGM\rho/2pR^2$, so if we use the same ideal gas law approximation, we obtain $\nu_{\max} \propto \nu_{ac} \propto M/(R^2T^{1/2})$. The maximum frequency is thus related to the fundamental stellar parameters (mass, radius, and temperature).

The proportionality relations for $\Delta\nu$ and $\nu_{\max}$ form the basis of scaling relations, asteroseismology's strongest weapon, which are obtained by rearranging the two equations above:

$$\left(\frac{M}{M_\odot}\right) = \left(\frac{\nu_{\max}}{\nu_{\max,\,\odot}}\right)^3 \left(\frac{\Delta\nu}{\Delta\nu_\odot}\right)^{-4} \left(\frac{T}{T_\odot}\right)^{3/2}, \tag{2.52}$$

$$\left(\frac{R}{R_\odot}\right) = \left(\frac{\nu_{\max}}{\nu_{\max,\,\odot}}\right) \left(\frac{\Delta\nu}{\Delta\nu_\odot}\right)^{-2} \left(\frac{T}{T_\odot}\right)^{1/2}. \tag{2.53}$$

Scaling relations provide fundamental parameters to ~10% accuracy (Huber et al. 2011, 2017). Tests are underway to compare fundamental parameters from eclipsing binary stars and asteroseismic scaling relations, which should give a definitive answer on the accuracy of scaling relations.

References

Aerts, C., Christensen-Dalsgaard, J., & Kurtz, D. W. 2010, Asteroseismology (Dordrecht: Springer)

Agafonov, M. I., Sharova, O. I., & Richards, M. T. 2009, ApJ, 690, 1730

Appenzeller, I. 2013, Introduction to Astronomical Spectroscopy (Cambridge: Cambridge Univ. Press)

Appourchaux, T., Belkacem, K., Broomhall, A.-M., et al. 2013, Introduction to Astronomical Spectroscopy (Cambridge: Cambridge Univ. Press)

Appourchaux, T., et al. 2010, A&ARv, 18, 197

Barnes, J. R., Collier Cameron, A., James, D. J., & Donati, J.-F. 2000, MNRAS, 314, 162

Baron, F. 2016, in Astronomy at High Angular Resolution, Vol. 439, ed. H. M. J. Boffin, et al. (Basel: Springer), 75

Bedding, T. R. 2014, in Asteroseismology, ed. P. L. Pallé, & C. Esteban (Cambridge: Cambridge Univ. Press), 60

Bedding, T. R., & Kjeldsen, H. 2003, PASA, 20, 203

Benson, J. A., Hutter, D. J., Johnston, K. J., et al. 2004, Proc. SPIE, 5491, 464

Berdyugin, A., Piirola, V., Sadegi, S., et al. 2016, A&A, 591, A92

Bessell, M. S. 1990, PASP, 102, 1181

Bessell, M., & Murphy, S. 2012, PASP, 124, 140

Bjorkman, K. S. 2012, in IAU Symp. 282, From Interacting Binaries to Exoplanets: Essential Modeling Tools, ed. M. T. Richards, & I. Hubeny (Cambridge: Cambridge Univ. Press), 173

Boffin, H. M. J., Hussain, G., Berger, J.-P., & Schmidtobreick, L. 2016, in Astronomy at High Angular Resolution (Basel: Springer), 439

Broomhall, A. -M., Chaplin, W. J., Davies, G. R., et al. 2009, MNRAS, 396, L100

Brown, A. G. A., Vallenari, A., Prusti, T., et al. 2016a, A&A, 595, A2

Brown, P. J., Breeveld, A., Roming, P. W. A., & Siegel, M. 2016b, AJ, 152, 102

Brown, T. M. 1991, ApJ, 371, 396

Casagrande, L., & VandenBerg, D. A. 2014, MNRAS, 444, 392

Che, X., Monnier, J. D., Tycner, C., et al. 2012, ApJ, 757, 29

Christensen-Dalsgaard, J. 2003, Lecture Notes on Stellar Oscillations http://users-phys.au.dk/jcd/oscilnotes/contents.html

Christensen-Dalsgaard, J., & Daeppen, W. 1992, A&ARv, 4, 267

Cousins, A. W. J. 1976, MmRAS, 81, 25
Cowling, T. G. 1941, MNRAS, 101, 367
Creech-Eakman, M. J., Romero, V., Westpfahl, D., et al. 2008, Proc. SPIE, 7013, 70130X
Crossfield, I. J. M. 2014, A&A, 566, A130
Deutsch, A. J. 1957, AJ, 62, 139
Eastman Kodak Company 2012, KAF-16801E 4096 (H) × 4096 (V) Pixel Performance Specification, Technical Report (Rochester, NY: Image Sensor Solutions)
Eastman, J., Siverd, R., & Gaudi, B. S. 2010, PASP, 122, 935
García, R. A., Régulo, C., Turck-Chiéze, S., et al. 2001, SoPh, 200, 361
Grindlay, J., Tang, S., Los, E., & Servillat, M. 2012, in IAU Symp. 285, New Horizons in Time Domain Astronomy, ed. E. Griffin, R. Hanisch, & R. Seaman (Cambridge: Cambridge Univ. Press), 29
Groote, D., Tuvikene, T., Edelmann, H., et al. 2014, in Astroplate 2014, ed. L. Mišková,, & S. Vítek (Prague: Institute of Chemical Technology), 53
Guzik, J. A., Kaye, A. B., Bradley, P. A., Cox, A. N., & Neuforge, C. 2000, ApJL, 542, L57
Hadrava, P. 2009, A&A, 494, 399
Haguenauer, P., Abuter, R., Alonso, J., et al. 2008, Proc. SPIE, 7013, 70130C
Hao, J. 1998, ApJ, 500, 440
Harris, H. C., Dahn, C. C., Zacharias, N., et al. 2016, AJ, 152, 118
Hatzes, A. P. 1988, PhD thesis, California Univ., Santa Cruz
Hendry, P. D., & Mochnacki, S. W. 2000, ApJ, 531, 467
Hobbs, G., & Dai, S. 2017, arXiv:1707.01615 [astro-ph.IM])
Högbom, J. A. 1974, A&AS, 15, 417
Huber, D., Bedding, T. R., Stello, D., et al. 2011, ApJ, 743, 143
Huber, D., Zinn, J., Bojsen-Hansen, M., et al. 2017, ApJ, 844, 102
Hummel, C. A., Carquillat, J.-M., Ginestet, N., et al. 2001, AJ, 121, 1623
Jankov, S., Cvetkovic, Z., & Pavlovic, R. 2014, SerAJ, 188, 1
Jiménez, A., García, R. A., & Pallé, P. L. 2011, ApJ, 743, 99
Johnson, H. L. 1965, ApJ, 141, 923
Jones, R. C. 1941, JOSA, 31, 488
Kippenhahn, R., & Weigert, A. 1994, Stellar Structure and Evolution (Basel: Springer)
Kirk, B., Conroy, K., & Prša, A. 2016, AJ, 151, 68
Kloppenborg, B., Stencel, R., & Monnier, J. D. 2010, Natur, 464, 870
Kolokolova, L., Hough, J., & Levasseur-Regourd, A.-C. 2015, Polarimetry of Stars and Planetary Systems (Cambridge: Cambridge Univ. Press)
Kovári, Z., Strassmeier, K. G., Granzer, T., et al. 2004, A&A, 417, 1047
Kumar, P., Ao, C. O., & Quataert, E. J. 1995, ApJ, 449, 294
Ledoux, P., Schwarzschild, M., & Spiegel, E. A. 1961, ApJ, 133, 184
Lindegren, L., & Dravins, D. 2003, A&A, 401, 1185
Manser, C. J., Gänsicke, B. T., Marsh, T. R., et al. 2016, MNRAS, 455, 4467
Marrese, P. M., Boschi, F., & Munari, U. 2003, A&A, 406, 995
Marsh, T. R., Horne, K., Schlegel, E. M., Honeycutt, R. K., & Kaitchuck, R. H. 1990, ApJ, 364, 637
Mazeh, T., Tsodikovich, Y., Segal, Y., et al. 2009, MNRAS, 399, 906
Milone, E. F. F., & Sterken, C. (ed.) 2011, Astronomical Photometry: Past, Present, and Future, in Astrophysics and Space Science Library, Vol. 373 (New York: Springer)

Monnier, J. D. 2003, RPPh, 66, 789
Morgan, W. W., Keenan, P. C., & Kellman, E. 1943, An Atlas of Stellar Spectra, with An Outline of Spectral Classification (Chicago, IL: Univ. Chicago Press)
Müller, H. 1948, JOSA, 38, 661
Munari, U. 1999, BaltA, 8, 73
Munari, U., & Tomasella, L. 1999, A&AS, 137, 521
Munari, U., Tomov, T., Zwitter, T., et al. 2001, A&A, 378, 477
Nitschelm, C. 1988, A&AS, 74, 67
Payne, C. H. 1925, PhD thesis, Radcliffe College
Perryman, M. A. C., Lindegren, L., Kovalevsky, J., et al. 1997, A&A, 323, L49
Piirola, V. 2010, in ASP Conf. Proc. 435, Binaries—Key to Comprehension of the Universe, ed. A. Prša, & M. Zejda (San Francisco, CA: ASP), 225
Pina, R. K., & Puetter, R. C. 1992, PASP, 104, 1096
Piskunov, N. E., Tuominen, I., & Vilhu, O. 1990, A&A, 230, 363
Piskunov, N. E., & Wehlau, W. H. 1990, A&A, 233, 497
Prša, A., Batalha, N., Slawson, R.W., et al. 2011, AJ, 141, 83
Readhead, A. C. S., Nakajima, T. S., Pearson, T. J., et al. 1988, AJ, 95, 1278
Rice, J. B., Wehlau, W. H., & Khokhlova, V. L. 1989, A&A, 208, 179
Rossiter, R. A. 1924, ApJ, 60, 15
Rucinski, S. M. 1992, AJ, 104, 1968
Saio, H. 1993, Ap&SS, 210, 61
Simon, K. P., & Sturm, E. 1994, A&A, 281, 286
Smart, W. M., & Green, E. b. R. M. 1977, Textbook on Spherical Astronomy (Cambridge: Cambridge Univ. Press)
Soffel, M., Klioner, S. A., Petit, G., et al. 2003, AJ, 126, 2687
Stokes, G. G. 1852, RSPT, 142, 463
Strassmeier, K. G., & Rice, J. B. 2003, A&A, 399, 315
Tassoul, M. 1980, ApJS, 43, 469
ten Brummelaar, T. A., McAlister, H. A., Ridgway, S. T., et al. 2005, ApJ, 628, 453
Thiebaut, E., & Giovannelli, J.-F. 2010, ISPM, 27, 97
Thompson, A. R., Moran, J. M., & Swenson, G. W. Jr. 2001, Interferometry and Synthesis in Radio Astronomy (2nd ed. New York: Wiley)
Thompson, S. E., Everett, M., Mullally, F., et al. 2012, ApJ, 753, 86
Tinbergen, J. 2005, Astronomical Polarimetry (Cambridge: Cambridge Univ. Press)
Tody, D. 1986, Proc. SPIE, 627, 733
Townsend, R., Teitler, S., & Paxton, B. 2014, in IAU Symp. 301, Precision Asteroseismology, ed. J. A. Guzik, et al. (Cambridge: Cambridge Univ. Press), 505
Tsvetkov, M. K., Stavrev, K. Y., Tsvetkova, K. P., Ivanov, P. V., & Iliev, M. S. 1994, in IAU Symp. 161, Astronomy from Wide-Field Imaging, ed. H. T. MacGillivray (Dordrecht: Springer), 359
van Altena, W. F. 2013, Astrometry for Astrophysics (Cambridge: Cambridge Univ. Press)
Vincent, A., Piskunov, N. E., & Tuominen, I. 1993, A&A, 278, 523
Vogt, S. S., & Penrod, G. D. 1983, PASP, 95, 565
Vogt, S. S., Penrod, G. D., & Hatzes, A. P. 1987, ApJ, 321, 496
Zucker, S., & Mazeh, T. 1994, ApJ, 420, 806
Zucker, S., Torres, G., & Mazeh, T. 1995, ApJ, 452, 863

Part II

Theoretical Background

Chapter 3

Geometry and Dynamics of Binary Stars

The description of binary star orbits is a celebrated problem of classical physics, discussed in most textbooks on analytical mechanics. However, the analytical solution assumes point masses, and it ignores any effects of tides or axial rotation. This is in clear violation of observations, and numerical approximations are necessary.

One of the most widely used approximations, the *Roche* model, is based on the principle of equipotential surfaces. It is built on the following assumptions:

1. stars in a binary system gravitationally interact as point masses that are surrounded by massless envelopes;
2. the damping timescales of any deviations from hydrostatic equilibrium are much shorter than the orbital period P, so the shape of the components is determined by the instantaneous force field; and
3. the star rotates about its axis as a rigid body, without any differential rotation.

The model was first formulated by the French astronomer Édouard Albert Roche (1820–1883) in 1849.

None of these three assumptions are fulfilled exactly, yet the deviations are reasonably small that the resulting model is still meritorious, mostly thanks to its simplicity.

3.1 A Quick Look at the Roche Model

The basic principles of the Roche model can be highlighted through a simplified treatment that invokes additional assumptions:

- orbits in a binary system are circular;
- rotational axes are aligned with the revolution axis; and
- the stars are synchronized so that the rotational angular velocity equals the orbital angular velocity.

doi:10.1088/978-0-7503-1287-5ch3

While these assumptions come at the expense of generality, a fair number of close systems (i.e., systems with orbital periods up to a few days) satisfy them, and it is instructive to briefly give up generality in favor of simplicity. We will abandon these additional assumptions shortly and derive a fully general expression for arbitrary Roche surfaces.

Let us start by showing that the equipotential lines of the effective gravitational potential fully determine the shapes of the stars. In the interior of either star, the net force exerted on a test particle m is given by Newton's law:

$$\boldsymbol{F}_p + \boldsymbol{F}_g + m\omega^2\tilde{\boldsymbol{r}} = 0, \tag{3.1}$$

where $\boldsymbol{F}_p$ is force of the pressure gradient, $\boldsymbol{F}_g$ is gravitational force, ω is angular velocity of the system, and $\tilde{\boldsymbol{r}}$ is the vector in the xy plane that originates at the axis of rotation, is perpendicular to it, and points to the test particle m (see Figure 3.1). Under the above assumptions, the particle will be at rest in the rotating coordinate system, i.e., independent of time. $\boldsymbol{F}_p$ may be written as the gradient of pressure:

$$\boldsymbol{F}_p = -V\nabla p, \tag{3.2}$$

and, because of the time independence, the remaining two terms in Equation (3.1) can be written as the gradient of an effective potential ψ:

$$\boldsymbol{F}_g + m\omega^2\tilde{\boldsymbol{r}} = -m\nabla\psi. \tag{3.3}$$

Combining Equations (3.1)–(3.3) yields:

$$-V\nabla p - m\nabla\psi = 0 \quad \Rightarrow \quad \nabla p = -\rho\nabla\psi. \tag{3.4}$$

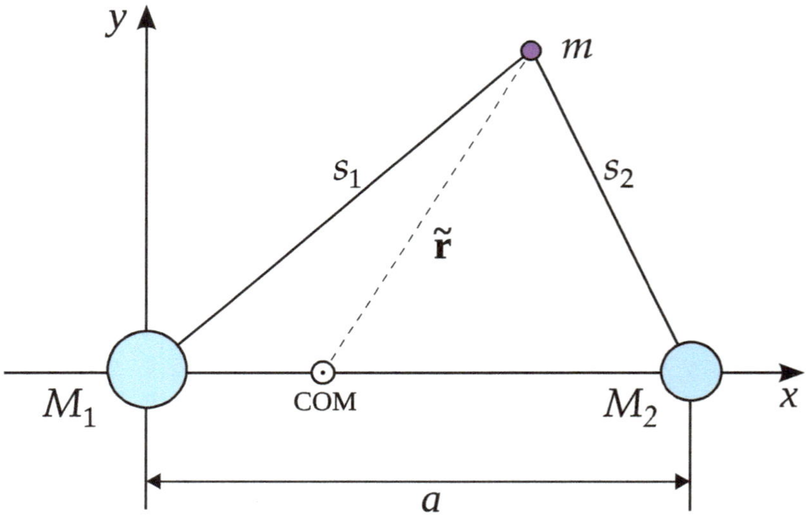

Figure 3.1. Binary system geometry. The origin of the coordinate system is in star M_1, not in the center of mass.

This implies that ∇p and $\nabla \psi$ are parallel. The density gradient $\nabla \rho$ is also parallel, which we can readily see by evaluating the curl of both sides in Equation (3.4):

$$\nabla \times \nabla p = -\nabla \times \rho \nabla \psi = -\rho \nabla \times \nabla \psi - \nabla \rho \times \nabla \psi. \tag{3.5}$$

Since the curl of a gradient is 0, it immediately follows that:

$$\nabla \rho \times \nabla \psi = 0 \quad \Rightarrow \quad \nabla \rho || \nabla \psi || \nabla p. \tag{3.6}$$

This implies that surfaces of constant potential, constant pressure, and constant density coincide. Once ψ is given, we can obtain the shape of the corresponding equipotential surface, and—according to Equation (3.6)—that equipotential surface also represents the physical boundary of the star.

We can also assess the extent of the second Roche model assumption, the dynamical timescale at which the star restores hydrostatic equilibrium, from the equation of motion for a spherical shell inside a spherically symmetric star:

$$\ddot{r} = -\frac{Gm}{r^2} - \frac{1}{\rho}\frac{dp}{dr}. \tag{3.7}$$

Say the pressure gradient vanishes; the shell will then start to freefall toward the center with an acceleration:

$$\ddot{r} \sim \frac{R}{\tau_{\rm ff}^2} \Rightarrow \tau_{\rm ff} = \sqrt{\frac{R}{\ddot{r}}} = \sqrt{\frac{R^3}{GM}} = \frac{1}{2}(\bar{\rho}G)^{-1/2}, \tag{3.8}$$

where $\bar{\rho} = 3M/4\pi R^3$ is the average density. For solar values we obtain $\tau_{\rm ff} \approx 1600$ s, or half an hour. While this is a crude estimate of the order of magnitude, asteroseismic results indicate that the typical timescales for re-establishing hydrostatic equilibrium are ~15 minutes.

Figure 3.1 depicts a typical binary system layout. M_1 and M_2 are point masses of individual stars, a is the separation between the masses, m is a test particle mass, and s_1 and s_2 are respective distances between m and M_1 and between m and M_2. The origin of the coordinate system is placed at M_1. The system revolves about the axis that goes through the barycenter and is parallel with the z-axis.

For the purpose of deriving a convenient form of the potential, we use spherical coordinates (Figure 3.2):

$$\begin{aligned} x &= r \sin\theta \cos\phi = \lambda r, \\ y &= r \sin\theta \sin\phi = \mu r, \\ z &= r \cos\theta \qquad\quad = \nu r, \end{aligned} \tag{3.9}$$

where λ, μ, and ν are direction cosines. Using this convention, the polar angle θ assumes values from 0° at $+z$ pole to 180° at $-z$ pole; the azimuthal angle ϕ assumes values from 0° to 360° in a positive (counterclockwise) sense, starting from the x-axis. The x-axis points toward the companion star (Figure 3.1). We will refer to angles θ and ϕ as colatitude and longitude, respectively.

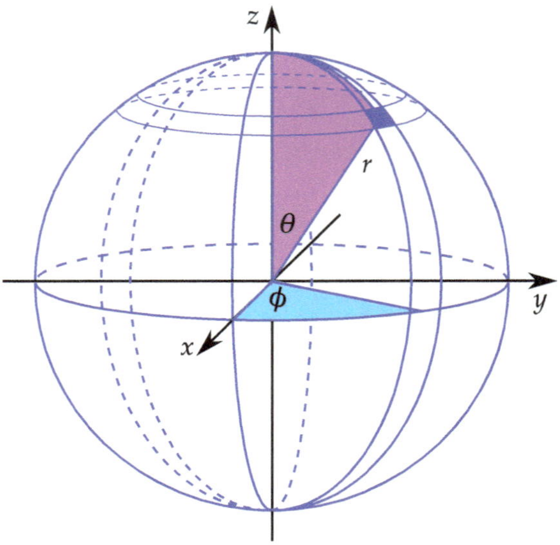

Figure 3.2. Spherical coordinate system convention: the polar angle θ spans values from 0° at $+z$ pole to 180° at $-z$ pole; the azimuth angle ϕ spans values from 0° to 360° in a positive (counterclockwise) direction, starting from the x-axis.

Effective potential ψ can be written out explicitly as:

$$\psi = -\frac{GM_1}{s_1} - \frac{GM_2}{s_2} - \frac{1}{2}\omega^2\tilde{r}^2. \tag{3.10}$$

Note that:

$$s_1 = r, \quad s_2 = \sqrt{r^2 - 2xa + a^2} \text{ and } \tilde{r} = \sqrt{(x - x_{bc})^2 + y^2}, \tag{3.11}$$

where x_{bc} is the barycenter of the system. It lies on the x-axis, and its position depends on the mass ratio $q = M_2/M_1$ of the binary:

$$x_{bc} = \frac{\sum_i M_i x_i}{\sum_i M_i} = \frac{qa}{1 + q}. \tag{3.12}$$

Equation (3.10) can thus be written as:

$$\begin{aligned}\psi(r, \lambda, \nu) = &-\frac{GM_1}{r} - \frac{GM_2}{\sqrt{r^2 - 2ar\lambda + a^2}} \\ &- \frac{1}{2}\omega^2\left(r^2(1 - \nu^2) - 2r\lambda x_{bc} + x_{bc}^2\right).\end{aligned} \tag{3.13}$$

Finally, we use Kepler's law $\omega^2 a^3 = G(M_1 + M_2)$ and Equation (3.12) to derive the full form of the effective potential:

$$\psi(r, \lambda, \nu) = -\frac{GM_1}{a}\left[\frac{a}{r} + q\left(\frac{a}{\sqrt{r^2 - 2ar\lambda + a^2}} - \frac{r\lambda}{a}\right) + \frac{1}{2}\frac{r^2}{a^2}(1 + q)(1 - \nu^2) + \frac{1}{2}\frac{q^2}{1 + q}\right]. \tag{3.14}$$

Following Kopal (1959), we can transform ψ to a dimensionless form:

$$\Omega = -\frac{a\psi}{GM_1} - \frac{1}{2}\frac{q^2}{1 + q}, \tag{3.15}$$

which, by substituting ψ with Equation (3.14) and introducing the dimensionless variable $\rho = r/a$, takes the following form:

$$\Omega = \frac{1}{\rho} + q\left(\frac{1}{\sqrt{\rho^2 - 2\rho\lambda + 1}} - \rho\lambda\right) + \frac{1}{2}(1 + q)(1 - \nu^2)\rho^2. \tag{3.16}$$

This form of the potential is frequently referred to as the *Kopal* potential. It corresponds to the effective gravitational potential of a binary under the assumptions laid out in the beginning of this section. Because it bounds the stars, we typically refer to it as the *surface* potential.

As the origin of the coordinate system is in M_1, the corresponding equipotential value around M_2 needs to be transformed. We do that by a symmetrical inversion of Ω and q:

$$\Omega' = \frac{\Omega}{q} + \frac{1}{2}\frac{q - 1}{q}, \quad q' = \frac{1}{q}. \tag{3.17}$$

Thus, to compute the potential of star M_2, we compute Ω as if it was star M_1, but then transform q and Ω using Equation (3.17). Note that Ω and Ω' apply only to circular orbits and aligned, synchronous rotation.

3.2 Roche Model Generalization

Abandoning the above assumptions is straightforward, but it needs a bit more legwork and care when deriving the generalized expressions. Let us first formalize and make explicitly clear the choices of coordinate systems.

- Let system S be an inertial coordinate system with the origin in the barycenter, the z-axis aligned with the axis of revolution, and the x-axis directed toward star M_2 at some reference time t_0. The basis vectors in this system are $\hat{\mathbf{i}}^*$, $\hat{\mathbf{j}}^*$, and $\hat{\mathbf{k}}^*$, and the components of $\boldsymbol{r}^*$ are (x^*, y^*, z^*).
- Let system S' be a noninertial coordinate system with the origin in star M_1, the z-axis aligned with the axis of rotation. The basis vectors in this system are $\hat{\mathbf{i}}'$, $\hat{\mathbf{j}}'$, and $\hat{\mathbf{k}}'$, and the components of vector $\boldsymbol{r}'$ are (x', y', z').

- Let system C be a canonical noninertial coordinate system with the origin in star M_1, the z-axis aligned with the axis of revolution, and the x-axis pointing toward star M_2 at all times. The basis vectors in this system are $\hat{\mathbf{\imath}}$, $\hat{\mathbf{\jmath}}$, and $\hat{\mathbf{k}}$, and the components of vector $\boldsymbol{r}$ are (x, y, z).

Figure 3.3 depicts these coordinate systems graphically. The rationale for introducing three separate coordinate systems is simple: (1) we need an inertial coordinate system S because an equation of motion can only be written in an inertial coordinate system, (2) we need a noninertial coordinate system S' because *relative* acceleration and velocity of a rigid body in that system will be 0, and (3) we use a noninertial coordinate system C as a matter of convention, to express the potential in terms of coordinates x, y, and z. The transformation of the acceleration vector from a noninertial to an inertial coordinate system is done by the following equation (Thornton & Marion 2004):

$$\ddot{\boldsymbol{r}}^* = \ddot{\boldsymbol{R}}^* + \ddot{\boldsymbol{r}}' + \dot{\boldsymbol{\omega}}'_{\text{rot}} \times \boldsymbol{r}' + \boldsymbol{\omega}'_{\text{rot}} \times \boldsymbol{\omega}'_{\text{rot}} \times \boldsymbol{r}' + 2\boldsymbol{\omega}'_{\text{rot}} \times \dot{\boldsymbol{r}}', \tag{3.18}$$

where $\boldsymbol{R}^*$ is a vector from the origin of S to the origin of S', the asterisk denotes vectors in an inertial coordinate system S, and the prime denotes vectors in a noninertial coordinate system S'. The benefit of choosing S' becomes immediately obvious: in that system $\ddot{\boldsymbol{r}}' = 0$ and $\dot{\boldsymbol{r}}' = 0$; in addition, as $\boldsymbol{\omega}'_{\text{rot}} = \text{const.}$, $\dot{\boldsymbol{\omega}}'_{\text{rot}} = 0$. Thus, Equation (3.18) reduces to:

$$\ddot{\boldsymbol{r}}^* = \ddot{\boldsymbol{R}}^* + \boldsymbol{\omega}'_{\text{rot}} \times \boldsymbol{\omega}'_{\text{rot}} \times \boldsymbol{r}'. \tag{3.19}$$

Acceleration $\ddot{\boldsymbol{R}}^*$ in an inertial coordinate system is only due to the gravitational force at the origin of S':

$$\ddot{\boldsymbol{R}}^* = \frac{GM_2}{d^2}\frac{\hat{\mathbf{\imath}}^*}{d}, \tag{3.20}$$

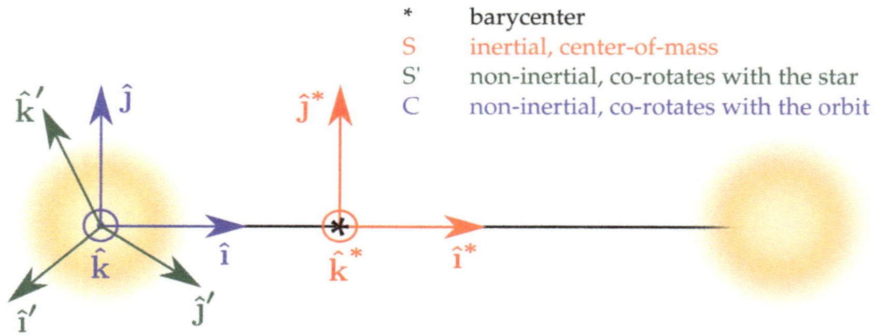

Figure 3.3. Coordinate systems in use for formulating the generalized Roche potential. System S is inertial, with basis vectors $\hat{\mathbf{\imath}}^*$, $\hat{\mathbf{\jmath}}^*$, $\hat{\mathbf{k}}^*$. System S' is noninertial, with basis vectors $\hat{\mathbf{\imath}}'$, $\hat{\mathbf{\jmath}}'$, $\hat{\mathbf{k}}'$. System C is noninertial, with basis vectors $\hat{\mathbf{\imath}}$, $\hat{\mathbf{\jmath}}$, $\hat{\mathbf{k}}$.

where d is the instantaneous distance to star M_2. The triple product $\boldsymbol{\omega}'_{\text{rot}} \times \boldsymbol{\omega}'_{\text{rot}} \times \boldsymbol{r}'$ can be simplified by using the chain rule, $\boldsymbol{a} \times \boldsymbol{b} \times \boldsymbol{c} = \boldsymbol{b}(\boldsymbol{a} \cdot \boldsymbol{c}) - \boldsymbol{c}(\boldsymbol{a} \cdot \boldsymbol{b})$:

$$\boldsymbol{\omega}'_{\text{rot}} \times \boldsymbol{\omega}'_{\text{rot}} \times \boldsymbol{r}' = \boldsymbol{\omega}'_{\text{rot}}(\boldsymbol{\omega}'_{\text{rot}} \cdot \boldsymbol{r}') - \omega'^2_{\text{rot}}\boldsymbol{r}', \tag{3.21}$$

and, as $\boldsymbol{\omega}'_{\text{rot}} = \omega'_{\text{rot}}\, \hat{\mathbf{k}}'$,

$$\boldsymbol{\omega}'_{\text{rot}} \times \boldsymbol{\omega}'_{\text{rot}} \times \boldsymbol{r}' = \omega'^2_{\text{rot}}[(\hat{\mathbf{k}}' \cdot \boldsymbol{r}')\hat{\mathbf{k}}' - \boldsymbol{r}']. \tag{3.22}$$

Acceleration in the inertial coordinate system S is then:

$$\ddot{\boldsymbol{r}}^* = \frac{GM_2}{d^2}\frac{\hat{\mathbf{i}}^*}{d} + \omega'^2_{\text{rot}}[(\hat{\mathbf{k}}' \cdot \boldsymbol{r}')\hat{\mathbf{k}}' - \boldsymbol{r}']. \tag{3.23}$$

The main takeaway of Equation (3.23) is that the right-hand side does not depend on time, i.e., there are no terms with $\dot{\boldsymbol{r}}'$ or $\ddot{\boldsymbol{r}}'$. Because of that, all vectors at a given instant in time are constant, and we can express them in *any* coordinate basis. As the basis of choice we use the canonical coordinate system C, and we denote vectors represented in that basis with $|_C$:

$$\ddot{\boldsymbol{r}}^* = \frac{GM_2}{d^2}\frac{\hat{\mathbf{i}}^*}{d}\bigg|_C + \omega'^2_{\text{rot}}[(\hat{\mathbf{k}}' \cdot \boldsymbol{r}')\hat{\mathbf{k}}' - \boldsymbol{r}']\bigg|_C. \tag{3.24}$$

In other words, vectors $\hat{\mathbf{i}}'$, $\hat{\mathbf{k}}'$, and $\boldsymbol{r}'$ are expressed in the basis of C, i.e., with unit vectors $\hat{\mathbf{i}}$, $\hat{\mathbf{j}}$, $\hat{\mathbf{k}}$ and coordinates x, y, z. Because the x-axes of systems S and C are aligned, $\hat{\mathbf{i}}^* \equiv \hat{\mathbf{i}}$, while $\hat{\mathbf{k}}'$ and $\boldsymbol{r}'$ can be expressed as a rotation $\mathcal{R}(\xi)\, \hat{\mathbf{k}}$ and $\mathcal{R}(\xi)\, \boldsymbol{r}'$ for a given misalignment angle ξ. As there are no time-dependent vectors, we can write the right-hand side as the gradient of the potential:

$$\ddot{\boldsymbol{r}}^* = \nabla_{r'}\left\{\frac{GM_2}{d^2}\hat{\mathbf{i}} \cdot \boldsymbol{r}' + \frac{1}{2}\omega'^2_{\text{rot}}[(\hat{\mathbf{k}}' \cdot \boldsymbol{r}')\hat{\mathbf{k}}' - \boldsymbol{r}']^2\right\}. \tag{3.25}$$

Following the equation of motion in an inertial coordinate system, the left-hand side of Equation (3.25) can be written as:

$$\ddot{\boldsymbol{r}}^* = -\nabla_{r'} U - \frac{1}{\rho}\nabla_{r'} p, \tag{3.26}$$

where U is the Keplerian two-body potential:

$$U = \frac{GM_1}{\|\boldsymbol{r}^* - \boldsymbol{r}_1^*\|} + \frac{GM_2}{\|\boldsymbol{r}^* - \boldsymbol{r}_2^*\|}. \tag{3.27}$$

Putting it all together,

$$\frac{1}{\rho}\nabla_{r'} p = -\nabla_{r'}\left\{U + \frac{GM_2}{d^2}\hat{\mathbf{i}} \cdot \boldsymbol{r}' + \frac{1}{2}\omega'^2_{\text{rot}}[(\hat{\mathbf{k}}' \cdot \boldsymbol{r}')\hat{\mathbf{k}}' - \boldsymbol{r}']^2\right\}. \tag{3.28}$$

The expression in curly braces (with the negative sign) is a generalized version of the effective potential ψ:

$$\psi = -U - \frac{GM_2}{d^2}\hat{\mathbf{i}} \cdot \boldsymbol{r}' - \frac{1}{2}\omega_{\rm rot}^{\prime 2}[(\hat{\mathbf{k}}' \cdot \boldsymbol{r}')\hat{\mathbf{k}}' - \boldsymbol{r}']^2. \quad (3.29)$$

In a special case where the axis of rotation and axis of revolution are aligned, $\hat{\mathbf{k}}' \equiv \hat{\mathbf{k}}^* \equiv \hat{\mathbf{k}}$, Equation (3.29) can be written as:

$$\psi = -\frac{GM_1}{s_1} - \frac{GM_2}{s_2} - \frac{GM_2}{d^2}x - \frac{1}{2}\omega_{\rm rot}^2(x^2 + y^2). \quad (3.30)$$

By repeating the process given by Equations (3.10)–(3.15), we can derive a generalized version of Equation (3.16), first proposed by Wilson (1979), valid for eccentric orbits and asynchronous rotation:

$$\Omega = \frac{1}{\rho} + q\left(\frac{1}{\sqrt{\delta^2 + \rho^2 - 2\rho\lambda\delta}} - \frac{\rho\lambda}{\delta^2}\right) + \frac{1}{2}F^2(1 + q)\rho^2(1 - \nu^2), \quad (3.31)$$

where $\delta = d/a$ is the fractional instantaneous separation and F is the synchronicity parameter, defined as the ratio between the rotational and orbital angular velocity:

$$F = \frac{\omega_{\rm rot}}{\omega_{\rm orb}}. \quad (3.32)$$

For as long as we neglect differential rotation, the synchronicity parameter is a single number for the whole star. Value $F = 1$ corresponds to a synchronous rotation in a circular orbit. In case of elliptical orbits the rotation tends to synchronize because of tidal interactions between the two components. In eccentric cases one can define pseudo-synchronous rotation, where the angular rotation rate of the star is equal to the orbital angular velocity *at periastron*. Following the angular momentum conservation, angular velocity ω at a given point on the orbit is given by the following expression (Hut 1981):

$$\omega(t) = \frac{1}{\delta^2}\left(\frac{2\pi}{P}\right)\sqrt{1 - e^2}, \quad (3.33)$$

where e is the orbital eccentricity. Since the synchronicity parameter is defined with respect to periastron, $\delta = 1 - e$, introducing this into Equation (3.33) yields that for pseudo-synchronous rotation the synchronicity parameter is given by:

$$F = \sqrt{\frac{1 + e}{(1 - e)^3}}. \quad (3.34)$$

The formalism of asynchronous rotation was first developed by Limber (1963); for observational tests on real binaries see, among many others, van Hamme & Wilson (1986).

Having elliptical orbits and asynchronous rotation effectively means that additional approximations need to be made for moving objects in the generalized potential, namely, that the now present Coriolis force is neglected. Limber (1963) provides a thorough discussion on error estimates due to this approximation.

Generalized Kopal potentials Ω and Ω' *uniquely* determine the shapes and sizes of individual stars in a binary system at any point in the orbit. Exactly how this is done is the topic of the next section.

3.2.1 Shapes and Sizes of Stars in a Binary System

In the previous section we have shown that the surfaces of constant pressure and density coincide with equipotential surfaces. This in turn implies that the stellar shape is fully determined by its effective potential Ω: given the value of Ω, Equation (3.31) has to be inverted to obtain the radius ρ as a function of directions λ and ν. This can be readily achieved by expressing the potential at the star's pole ($\theta = 0$, thus $\lambda = 0$, $\nu = 1$):

$$\Omega = \frac{1}{\rho_{\text{pole}}} + q\left(\frac{1}{\sqrt{\delta^2 + \rho_{\text{pole}}^2}}\right), \tag{3.35}$$

and imposing that Ω is the same for any λ and ν,

$$\begin{aligned} \frac{1}{\rho} + q\left(\frac{1}{\sqrt{\delta^2 + \rho^2 - 2\rho\lambda\delta}} - \frac{\rho\lambda}{\delta^2}\right) + \frac{1}{2}F^2(1+q)\rho^2(1-\nu^2) \\ = \frac{1}{\rho_{\text{pole}}} + q\left(\frac{1}{\sqrt{\delta^2 + \rho_{\text{pole}}^2}}\right). \end{aligned} \tag{3.36}$$

Solving this equation iteratively for each λ and ν determines the shape and the size of the star. Figure 3.4 depicts five color-coded examples of an xz cross section of stellar shapes that correspond to the constant mass ratio $q = 0.5$ and different values of Ω. The shape of the star is determined by the equipotential it fills, so the value of potential Ω determines the morphology of the binary. The most important equipotential on Figure 3.4 is drawn in black and corresponds to the value $\Omega = 2.87585$: both stars have a teardrop shape, and they are in contact at only one point. This particular equipotential is referred to as the *Roche lobe*, and it plays a significant role in morphological classification of close binaries.

The amount of distortion of the stellar surface depends on the value of effective potential Ω. The deviation from a spherical form also causes nonuniform distribution of surface brightness. To quantify it, we need to relate a local surface element normal $\hat{e}_{\text{n}}$ with the unit radius vector $\hat{e}_r$. Figure 3.5 shows a semi-detached, lobe-filling component with a distorted surface element. Since the gradient of the

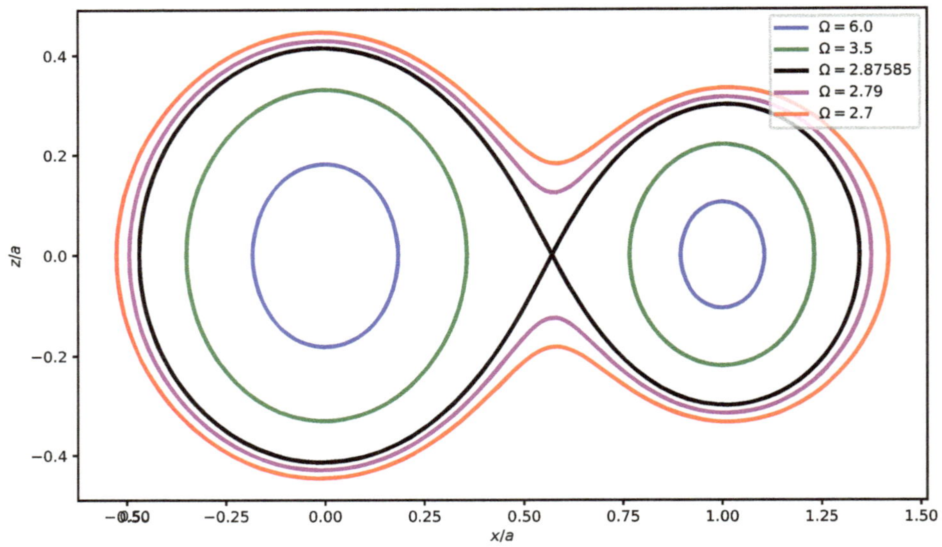

Figure 3.4. The xz cross section of the Roche coordinate system. Five color-coded examples are all depicted for the mass ratio $q = M_2/M_1 = 0.5$. The blue and green equipotentials correspond to a detached configuration; the black equipotential corresponds to the critical Roche lobe, and magenta and red equipotentials correspond to a contact configuration.

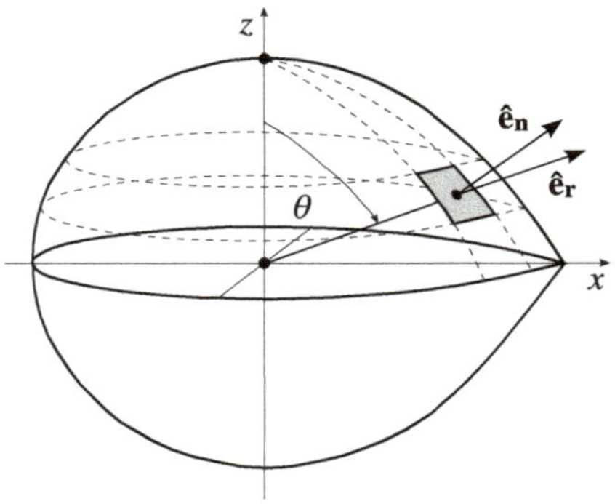

Figure 3.5. Surface distortion of a lobe-filling component. Unit vectors $\hat{e}_r$ and $\hat{e}_n$ enclose an angle γ that increases with distortion of the stellar surface.

potential is always perpendicular to equipotential surfaces and it points inward, the surface element normal may be written as:

$$\hat{e}_{\mathrm{n}} = -\frac{\nabla\Omega(\lambda, \mu, \nu)}{\|\nabla\Omega(\lambda, \mu, \nu)\|}, \tag{3.37}$$

where λ, μ, and ν are the direction cosines as defined by Equation (3.9). The unit radius vector is given by:

$$\hat{e}_r = (\lambda, \mu, \nu). \tag{3.38}$$

The angle γ between the two vectors is given by the dot product,

$$\cos\gamma = \hat{e}_{\mathbf{n}} \cdot \hat{e}_r = -\|\nabla\Omega(\lambda, \mu, \nu)\|^{-1}\left(\lambda\frac{\partial\Omega}{\partial x} + \mu\frac{\partial\Omega}{\partial y} + \nu\frac{\partial\Omega}{\partial z}\right), \tag{3.39}$$

where the gradient magnitude $\|\nabla\Omega(\lambda, \mu, \nu)\|$ is given by:

$$\|\nabla\Omega(\lambda, \mu, \nu)\| = \sqrt{\left(\frac{\partial\Omega}{\partial x}\right)^2 + \left(\frac{\partial\Omega}{\partial y}\right)^2 + \left(\frac{\partial\Omega}{\partial z}\right)^2}, \tag{3.40}$$

and individual partial derivatives of the effective potential follow from Equation (3.31):

$$\begin{aligned}
\frac{\partial\Omega}{\partial x} &= -\frac{x}{(x^2+y^2+z^2)^{3/2}} + \frac{q(d-x)}{([d-x]^2+y^2+z^2)^{3/2}} + F^2(1+q)x - \frac{q}{d^2},\\
\frac{\partial\Omega}{\partial y} &= -y\left[\frac{1}{(x^2+y^2+z^2)^{3/2}} + \frac{q}{([d-x]^2+y^2+z^2)^{3/2}} - F^2(1+q)\right],\\
\frac{\partial\Omega}{\partial z} &= -z\left[\frac{1}{(x^2+y^2+z^2)^{3/2}} + \frac{q}{([d-x]^2+y^2+z^2)^{3/2}}\right].
\end{aligned} \tag{3.41}$$

This allows us to explicitly write the surface element $d\sigma$:

$$d\sigma(\lambda, \mu, \nu) = \frac{\rho^2(\lambda, \mu, \nu)\sin\theta}{\cos\gamma(\lambda, \mu, \nu)}d\theta d\phi. \tag{3.42}$$

Integrating this equation over all surface elements yields the total surface of the distorted star:

$$S = \int_{\partial V} d\sigma(\lambda, \mu, \nu) = \int_0^{2\pi}\int_0^{\pi}\frac{\rho^2(\theta, \phi)\sin\theta}{\cos\gamma(\theta, \phi)}d\theta d\phi. \tag{3.43}$$

In a similar fashion, the volume of the star may be obtained by integrating over all volume elements:

$$V = \int_V dV(\lambda, \mu, \nu) = \int_0^{2\pi}\int_0^{\pi}\int_0^{\rho(\theta, \phi)} \rho^2(\theta, \phi)\sin\theta dr d\theta d\phi. \tag{3.44}$$

3.2.2 Lagrangian Points

Assuming aligned, circular motion, in a rotating two-body coordinate frame such as the one depicted in Figure 3.1, there are five points in which the net force exerted on a particle mass is 0. These points, shown in Figure 3.6, are called *Lagrangian* points. They satisfy the following condition:

$$\boldsymbol{F}_{\text{eff}} = -m\nabla\psi = 0. \tag{3.45}$$

Writing out ψ and evaluating its derivative along z, the above condition becomes:

$$\boldsymbol{F}_{\text{eff},\, z} = \left(\frac{GM_1{}^3}{s_1} + \frac{GM_2{}^3}{s_2}\right)z = 0. \tag{3.46}$$

As the term in parentheses is always >0, it immediately follows that $z = 0$. Thus, all Lagrangian points are coplanar, lying in the orbital (xy) plane. Equation (3.45) cannot be solved analytically, but there are useful approximations (Taff 1985) accurate to 10^{-6} that may be used in practice:

$$\begin{aligned} x(L_1) &= \zeta - \frac{1}{3}\zeta^2 - \frac{1}{9}\zeta^3 + \frac{58}{81}\zeta^4, \\ x(L_2) &= \zeta + \frac{1}{3}\zeta^2 - \frac{1}{9}\zeta^3 + \frac{50}{81}\zeta^4, \\ x(L_3) &= 1 - \frac{7}{12}\mu - \frac{1127}{20736}\mu^3 - \frac{7889}{248832}\mu^4, \end{aligned} \tag{3.47}$$

where $\mu = M_2/(M_1 + M_2)$ and $\zeta = (\mu/3)^{1/3}$.

Points L_1, L_2, and L_3 are located on the x-axis and are always unstable; points L_4 and L_5 are displaced in the $\pm y$-direction and may be either stable or unstable, depending on the ratio $M_2/(M_1 + M_2)$. The condition for the stability of equilibrium

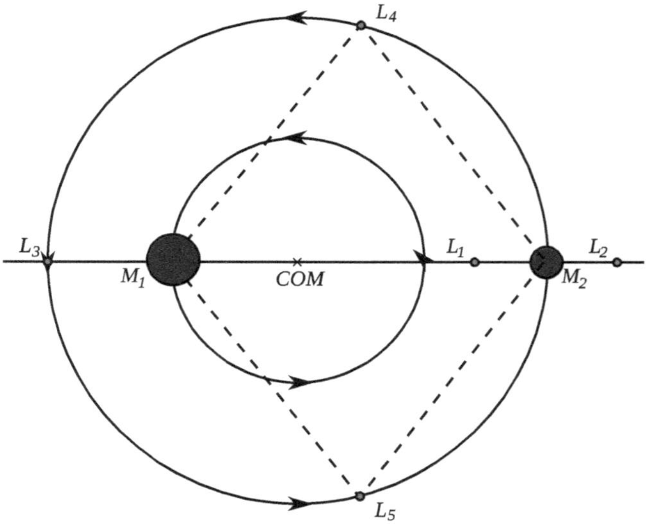

Figure 3.6. Lagrangian points in a rotating two-body system. The first three points (L_1 through L_3) lie on the straight line that passes through the center of the two stars. The remaining two points (L_4 and L_5) form equilateral triangles with the two stars.

points can be assessed by computing the response to a slight perturbation in position: if the resulting force points back to the equilibrium point, that point is stable; otherwise, it is unstable. As all Lagrangian points lie in the xy plane, let us first investigate the response to a displacement along the z-direction:

$$\left.\frac{\partial^2\psi}{\partial z^2}\right|_{z=0} = \frac{GM_1}{s_1^3} + \frac{GM_2}{s_2^3} > 0. \tag{3.48}$$

Thus, all Lagrangian points are stable to perturbation along z. To investigate perturbations along x and y, we introduce the following displacement:

$$x = x_0 + \delta x, \quad y = y_0 + \delta y, \quad z = 0, \tag{3.49}$$

where $(x_0, y_0, 0)$ are the coordinates of the Lagrangian point in question. The potential at each Lagrangian point can then be expanded to second order as:

$$\begin{aligned}\psi = \psi_0 &+ \left.\frac{\partial\psi}{\partial x}\right|_{(x_0,\, y_0,\, 0)} \delta x + \left.\frac{\partial\psi}{\partial y}\right|_{(x_0,\, y_0,\, 0)} \delta y + \frac{1}{2}\left.\frac{\partial^2\psi}{\partial x^2}\right|_{(x_0,\, y_0,\, 0)} \delta x^2 \\ &+ \frac{1}{2}\left.\frac{\partial^2\psi}{\partial x\partial y}\right|_{(x_0,\, y_0,\, 0)} \delta x\delta y + \frac{1}{2}\left.\frac{\partial^2\psi}{\partial y^2}\right|_{(x_0,\, y_0,\, 0)} \delta y^2 + \dots .\end{aligned} \tag{3.50}$$

The equations of motion for the test particle along x and y are:

$$\ddot{x} - 2\omega\dot{y} = -\frac{\partial\psi}{\partial x}, \quad \ddot{y} + 2\omega\dot{x} = -\frac{\partial\psi}{\partial y}. \tag{3.51}$$

Putting Equations (3.49)–(3.51) together, we obtain the following system of equations:

$$\begin{aligned}\delta\ddot{x} - 2\omega\delta\dot{y} &= -\frac{\partial^2\psi}{\partial x^2}\delta x - \frac{\partial^2\psi}{\partial x\partial y}\delta y, \\ \delta\ddot{y} + 2\omega\delta\dot{x} &= -\frac{\partial^2\psi}{\partial x\partial y}\delta x - \frac{\partial^2\psi}{\partial y^2}\delta y.\end{aligned} \tag{3.52}$$

The solution for this system can be found by guesswork:

$$\delta x = \delta x_0 e^{\gamma t}, \quad \delta y = \delta y_0 e^{\gamma t}. \tag{3.53}$$

Plugging these into Equation (3.52) leaves us with the following system:

$$\begin{aligned}\left(\gamma^2 + \frac{\partial^2\psi}{\partial x^2}\right)\delta x_0 + \left(-2\gamma\omega + \frac{\partial^2\psi}{\partial x\partial y}\right)\delta y_0 &= 0, \\ \left(2\gamma\omega + \frac{\partial^2\psi}{\partial x\partial y}\right)\delta x_0 + \left(\gamma^2 + \frac{\partial^2\psi}{\partial y^2}\right)\delta y_0 &= 0.\end{aligned} \tag{3.54}$$

This system will have a nontrivial solution when the determinant equals 0:

$$\gamma^4 + \gamma^2\left(\frac{\partial^2\psi}{\partial x^2} + \frac{\partial^2\psi}{\partial y^2} + 4\omega^2\right) + \frac{\partial^2\psi}{\partial x^2}\frac{\partial^2\psi}{\partial y^2} - \left(\frac{\partial^2\psi}{\partial x\partial y}\right)^2 = 0. \tag{3.55}$$

The partial derivatives above expand to:

$$\begin{aligned}
\frac{\partial^2\psi}{\partial x^2} &= -GM_1\left(\frac{3x^2}{s_1^5} - \frac{1}{s_1^3}\right) - GM_2\left(\frac{3(x-a)^2}{s_2^5} - \frac{1}{s_2^3}\right) - \omega^2,\\
\frac{\partial^2\psi}{\partial y^2} &= -GM_1\left(\frac{3y^2}{s_1^5} - \frac{1}{s_1^3}\right) - GM_2\left(\frac{3y^2}{s_2^5} - \frac{1}{s_2^3}\right) - \omega^2,\\
\frac{\partial^2\psi}{\partial x\partial y} &= -\frac{3GM_1xy}{s_1^5} - \frac{3GM_2(x-a)y}{s_2^5}.
\end{aligned} \tag{3.56}$$

Let us define four convenience parameters:

$$\begin{aligned}
A &= \frac{GM_1}{s_1^3} + \frac{GM_2}{s_2^3},\\
B &= 3\left(\frac{GM_1}{s_1^5} + \frac{GM_2}{s_2^5}\right)y^2,\\
C &= 3\left(\frac{GM_1xy}{s_1^5} + \frac{GM_2(x-a)}{s_2^5}\right)y,\\
D &= 3\left(\frac{GM_1x^2}{s_1^5} + \frac{GM_2(x-a)^2}{s_2^5}\right).
\end{aligned} \tag{3.57}$$

Equations (3.56) then simplify into:

$$\frac{\partial^2\psi}{\partial x^2} = A - D - \omega^2, \quad \frac{\partial^2\psi}{\partial y^2} = A - B - \omega^2, \quad \frac{\partial^2\psi}{\partial x\partial y} = -C. \tag{3.58}$$

Because Lagrangian points L_1, L_2, and L_3 lie on the x-axis, it immediately follows that $B = C = 0$ and, as $s_1^2 \equiv x^2$ and $s_2^2 \equiv (x-a)^2$, $D = 3A$. Equation (3.55) then becomes:

$$\gamma^4 + \gamma^2(2\omega^2 - A) - 2A^2 + A\omega^2 + \omega^4 = 0, \tag{3.59}$$

with the following four solutions:

$$\gamma = \pm\sqrt{\frac{A - 2\omega^2 \pm [A(9A - 8\omega^2)]^{1/2}}{2}}. \tag{3.60}$$

If a Lagrangian point is to be stable to a perturbation, the coefficient A must be such that *all* γ are imaginary, so $A(9A - 8\omega^2)$ has to be real and negative. That happens when:

$$\frac{8}{9}\omega^2 \leqslant A \leqslant \omega^2. \tag{3.61}$$

Yet it turns out that, for all three Lagrangian points, A is always larger than ω^2. To see this, write A for the 1D case:

$$A = \frac{GM_1}{x^3} + \frac{GM_2}{(a-x)^3} = \frac{GM_1}{a^3}\left(\frac{a^3}{x^3} + \frac{qa^3}{(a-x)^3}\right). \tag{3.62}$$

Solve for the term GM_1/a^3 from Kepler's third law,

$$A = \omega^2 \frac{1}{1+q}\left(\frac{a^3}{x^3} + \frac{qa^3}{(a-x)^3}\right), \tag{3.63}$$

and evaluate $dA/dq = 0$; the minimum happens at $x = a/2$, i.e., $q = 1$ and $A = 8\omega^2$. In consequence, Lagrangian points L_1, L_2, and L_3 are *always* unstable.

Now let us consider Lagrangian points L_4 and L_5. Their positions are determined by an equilateral triangle, so $s_1 = s_2 = a$. Redoing the above exercise, we arrive at the stability criterion:

$$\frac{M_2}{M_1 + M_2} < 0.0385. \tag{3.64}$$

In other words, the limiting mass ratio $q \equiv M_2/M_1 \approx 0.04004$ impacts the stability of the Lagrangian points L_4 and L_5: if the mass ratio is smaller, the points are stable; otherwise, they are unstable.

3.2.3 Validity

Before we proceed, let us digress and discuss the validity of Equation (3.31). The $1/s$ dependence of the gravitational potential ψ is exact only if the star is spherically symmetric. In reality, however, stars are distorted by tidal interactions and rotation, and their mass distribution is no longer spherically symmetric. Chandrasekhar (1933) showed that the distortion of a binary system component is the same as if it rotated like a rigid body about its own axis with an angular velocity ω and then was tidally distorted by the secondary at a distance a from its center of gravity, the two effects being simply added. This superposition theorem is true to (and including) the fifth order in $\langle R\rangle/s$, where $\langle R\rangle$ is the mean stellar radius. To see this, let us add a quadrupole term to the gravitational potential of the secondary star:

$$\psi_2^G = -G\left(\frac{M_2}{s_2} + \frac{J_x + J_y + J_z - 3I}{2s_2^3} + \ldots\right), \tag{3.65}$$

where J_x, J_y, and J_z are the principal components of inertia of the secondary star at its center of gravity and I is the moment of inertia about the radius vector. These moments of inertia are of the order of $M_2\langle R_2\rangle^2$, and tidal interaction is of the order of $\langle R_2/s_2\rangle^3$, which yields the second term in the equation above of the order of $M_2/\langle R_2\rangle(\langle R_2\rangle/s_2)^6$.

The potential ψ_2^G in the vicinity of the primary star may be expanded in terms of Legendre polynomials up to the fifth order (Sterne 1939):

$$\psi_2^G = \frac{GM_2}{a^2} s_1 \cos\varphi + \frac{GM_2}{a} \sum_{i=2}^{4} \left(\frac{s_1}{a}\right)^i P_i(\cos\varphi) + \text{const}, \tag{3.66}$$

where φ is the angle between the x-axis and $\boldsymbol{r}$. The first term on the right describes the orbital acceleration of the primary star as a whole; as the origin of the corotating coordinate system is in the primary star, this term vanishes, as the center of gravity of the primary will always be at the origin. The remaining terms ($2 \leqslant i \leqslant 4$) constitute a tidal potential that causes the deviation from a spherically symmetric potential. To solve this analytically, we would need to assume a density distribution model for the star. Irrespective of the model, though, one thing is clear: the bulging causes *torque* in the system that is not accounted for by the point-mass approximation of the Roche model. This carries significant implications, and we return to this point when we discuss apsidal motion in Section 3.3.

3.2.4 Morphological Classification of Close Binaries

Morphology of binary stars is very important because it strongly influences both the evolution and observed appearance of the system. As we have shown in the previous section, the shapes of stars in a binary system depend solely on the value of Kopal potentials Ω and Ω'. Depending on their value, compared to the critical value of the Roche lobe, close binaries are classified into three distinct groups:

DETACHED BINARIES are systems in which both stars are bound within separate equipotential surfaces. Their evolution is largely independent from one another; that is why detached binaries are frequently used as physical laboratories for studying the properties of individual stars.

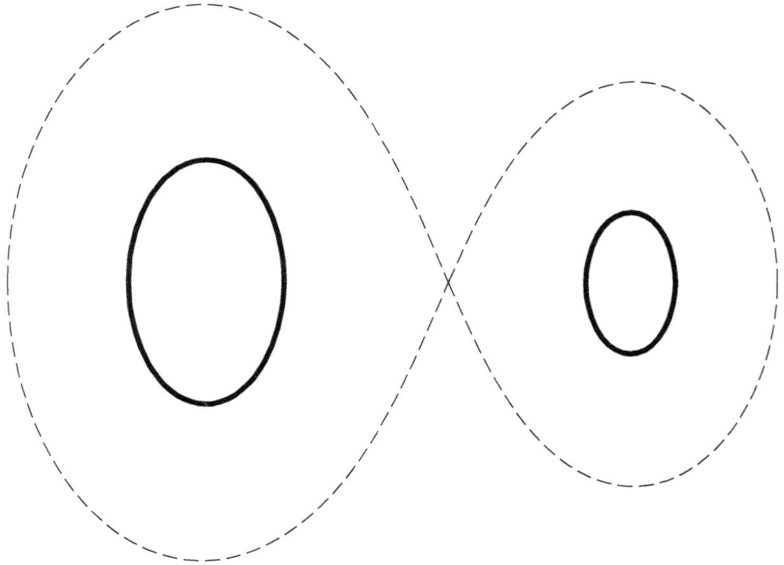

SEMI-DETACHED BINARIES are systems in which one star fills the Roche lobe and the other is bound within a separate equipotential surface. Semi-detached binaries frequently exhibit mass transfer, accretion disks, outbursts, hot spots, and similar transient phenomena.

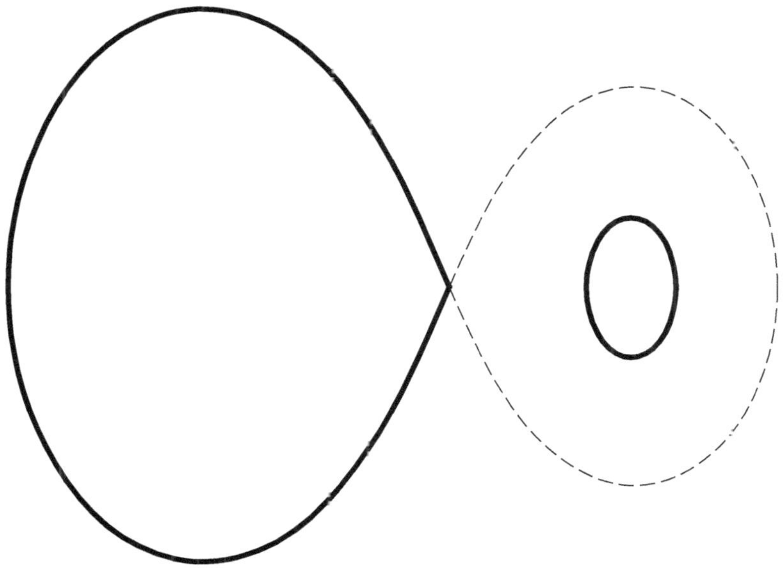

CONTACT BINARIES are systems in which both stars overflow the Roche lobe and share a common envelope. They may or may not be in a thermal equilibrium. Their evolution is closely coupled.

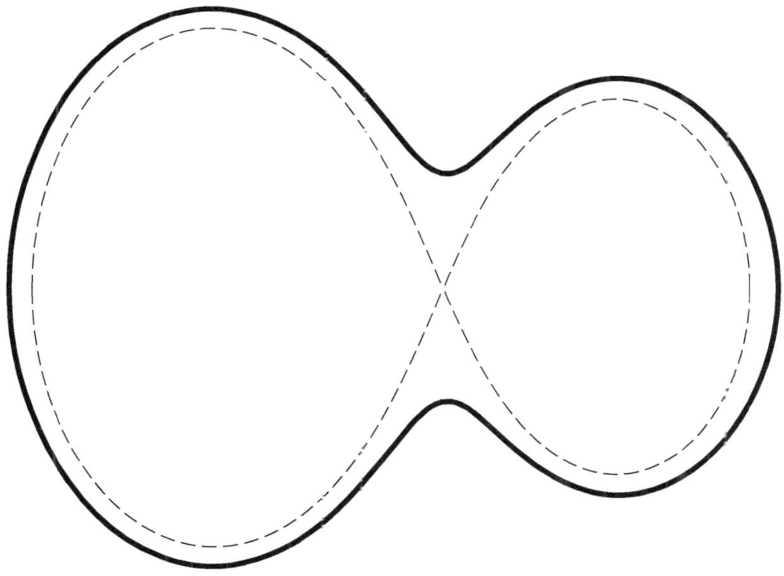

The term "overcontact" is sometimes used in the literature to denote this morphology class; here and throughout the book we use the term "contact" because it avoids confusion with the outer lobe overflow and mass loss through L_2 and/or L_3 points, which is what "overcontact" might imply. See Rucinski (1997) and Wilson (2001) for an in-depth discussion of this issue.

The Roche lobe is thus the limiting equipotential surface that distinguishes between different morphologies of close binary stars. The contact point of the Roche lobe is the *Lagrangian point* L_1. This point can be readily computed: either numerically or via the approximation given in Equation (3.6). The value of the potential in L_1, Ω_{L_1}, can then be computed by using Equation (3.31), where $\rho = x_{L_1}$. In the same manner, Ω_{L_2} may be obtained from the position x_{L_2}. These are inner and outer critical values of the potential: if the potential is larger than the inner potential, the star will not overflow the Roche lobe; if the potential is smaller than the outer potential, the system will experience mass loss through L_2 into space on a dynamical timescale. These critical potentials can be used to define the *fillout factor* of the system:

$$FF = \frac{\Omega - \Omega_{L_1}}{\Omega_{L_2} - \Omega_{L_1}}. \tag{3.67}$$

Thus, $FF = 0$ corresponds to the semi-detached case, $FF < 0$ to the detached case, and $FF \in (0, 1)$ to the contact case.

The classification of all binary stars into detached, semi-detached, and contact systems is of course very crude, as it provides no information beyond this most basic distinction. It would be far more useful if these discrete categories can be replaced by a continuous measure of "detachedness," where low values correspond to well-detached binaries and high values correspond to contact binaries. Such a measure, called the *morphology parameter*, was first proposed by Matijevič et al. (2012) and has been applied to the *Kepler* data set. The proposed classification is built on the method of locally linear embedding (LLE; Roweis & Saul 2000; Vander Plas & Connolly 2010), a dimensionality reduction tool that down-projects phased light curves onto a 2D manifold in a series of three steps. First, the method identifies nearest neighbors in the high-dimensional hyperspace spun by all observed light curves—in effect, this step groups light curves by similarity in the high-dimensional hyperspace. Next, the method characterizes the local geometry of each light curve by a linear combination of its neighbors, where coefficients of the linear combination are optimized and stored. Finally, the data are down-projected (mapped) onto a low-dimensional manifold (in this case, 2D) under the assumption that neighbors in the high-dimensional hyperspace will remain neighbors in the down-projected low-dimensional hyperspace. The algorithm is computationally inexpensive and depends on a single parameter, the number of nearest neighbors.

Figure 3.7 depicts a 2D manifold of all *Kepler* eclipsing binary light curves. The manifold is fit by a spline that captures the variation in morphology from well detached (left) to contact/ellipsoidal (right). The bottom panel of Figure 3.7 shows 15 representative light curves along the manifold. The depicted light curves are analytical approximations computed by `polyfit` (Prša et al. 2008) to form polynomial chains.

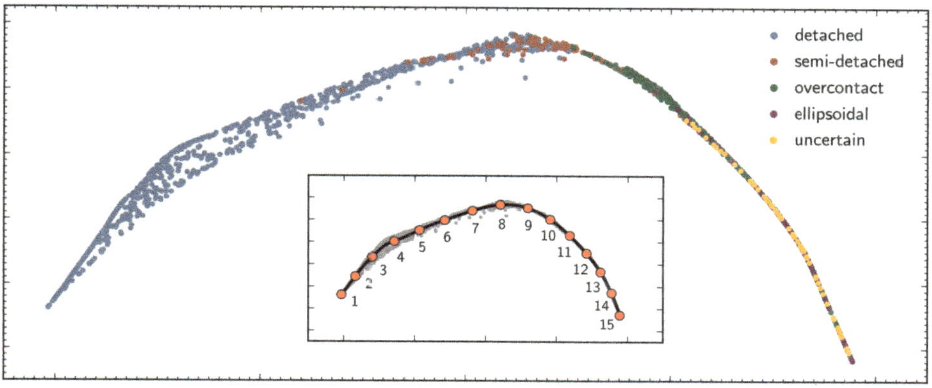

Figure 3.7. 2D manifold of all *Kepler* eclipsing binary light curves. Color-coding represents the manually classified light curves from Prša et al. (2011). The axes bear no physical significance and are thus not labeled—the scale and extent depend on the down-projection parameters (for details, see Matijevič et al. 2012). The inset depicts a spline fit to the manifold. The light curves are analytical `polyfit` representations that approximate the observed light curves.

Another recent classification application to eclipsing binary stars, by Kirk et al. (2016) for *Kepler* and by Kochoska et al. (2017) for *Gaia*, employed the t-distributed stochastic neighbor embedding (t-SNE; Van Der Maaten and Hinton 2008). The t-SNE method is primarily a visualization method where light curves are grouped by similarity and down-projected to spatially distinct manifolds. These manifolds are then automatically detected as clusters by using the DBSCAN algorithm (Ester et al. 1996). The main appeal of t-SNE is its ability to overcome the so-called "crowding" problem in other dimensionality reduction techniques. It depends on a single input parameter, *perplexity*. It instructs the method on how conservatively to group the light curves: it determines the number of nearest neighbors based on the local data density. Lower perplexity values result in a large number of small clusters, while higher perplexity values embed all light curves into a single large cluster. Figure 3.8 depicts the 2D manifold of clusters for a selection of simulated *Gaia* eclipsing binary light curves. The t-SNE/DBSCAN application resulted in nine distinct clusters, the representative light curves of which are shown in individual panels.

3.2.5 Critical Rotation

We digress here for a moment to point out another benefit of equipotential formulation of stellar shapes. Equipotential surfaces help us determine the breakup angular velocity ω_{crit}. Let us limit our discussion to a single star for a moment; a critically rotating star will be delimited by a single equipotential. The pole and the point radius will both lie on that equipotential:

$$\frac{1}{\rho_{\text{pole}}} = \frac{1}{\rho_{\text{point}}} + \frac{1}{2GM}\omega_{\text{crit}}^2\rho_{\text{point}}^2. \tag{3.68}$$

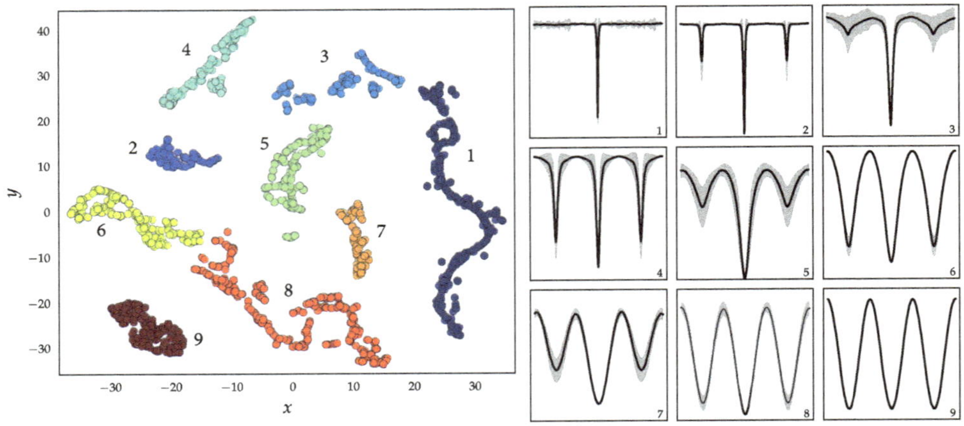

Figure 3.8. Left: t-SNE/DBSCAN results for the set of simulated *Gaia* eclipsing binary light curves. There are nine identified clusters that effectively classify the morphology. Right: average of all light curves in each cluster (black) and the 1σ region around the computed mean (gray region). Adopted from Kochoska et al. (2017). Credit: Kochoska et al., A&A, 206, A110, 2017, reproduced with permission © ESO.

At the breakup point, gravity and centrifugal forces are equal:

$$\frac{GM}{\rho_{\text{point}}^2} = \omega_{\text{crit}}^2 \rho_{\text{point}} \quad \Rightarrow \quad \omega_{\text{crit}}^2 = \frac{GM}{\rho_{\text{point}}^3}. \tag{3.69}$$

Plugging in ω_{crit} from Equation (3.69) to Equation (3.68) gives:

$$\frac{\rho_{\text{point}}}{\rho_{\text{pole}}} = 1 + \frac{1}{2GM}\omega_{\text{crit}}^2 \rho_{\text{point}}^3 = 1 + \frac{1}{2} = \frac{3}{2}. \tag{3.70}$$

Thus, the equatorial-to-polar radius ratio for a critical rotator is 3/2. The equatorial breakup velocity is then:

$$v_{\text{crit}}^2 = (\omega_{\text{crit}} \rho_{\text{point}})^2 = \frac{GM}{\rho_{\text{point}}} \equiv \frac{2GM}{3\rho_{\text{pole}}}. \tag{3.71}$$

If we introduce $\omega_{\text{rot}} \equiv \omega/\omega_{\text{crit}}$, we can express the angular velocity of the star as:

$$\omega = \omega_{\text{rot}} \omega_{\text{crit}} = \frac{GM\omega_{\text{rot}}}{\rho_{\text{point}}^3} = \frac{8GM\omega_{\text{rot}}}{27\rho_{\text{pole}}^3}. \tag{3.72}$$

3.3 Orbital Properties of Close Binaries

Owing to the Roche approximation of point masses, the motion of stars in a binary system around the barycenter constitutes a classical two-body problem. The analytical solution for a two-body problem can be found in most classical mechanics books (e.g., Goldstein et al. 2002; Thornton & Marion 2004), so here we will only recap the most important results.

A general solution of the two-body problem is elegantly solved by reducing[1] the problem of two bodies with masses M_1 and M_2 moving about the barycenter to the problem of a single, fictitious body with mass $\mu = M_1M_2/(M_1 + M_2)$ moving about the barycenter. We start out by writing a Hamiltonian:

$$H = \frac{1}{2}\mu\dot{r}^2 - \frac{GM_1M_2}{r}, \tag{3.73}$$

where $v \equiv \dot{r}$ is the orbital speed of the fictitious body that equals the relative speed between the two bodies. This is a separable first-order differential equation, but the solution for $r(t)$ is not analytical. That, of course, does not mean that we cannot integrate it either numerically or by using special functions (i.e., Hansen coefficients; Mardling 2013), but the usual route is to solve for $r(\upsilon)$ instead, where υ is the angle measured from periastron. The result is a well-known *conic section*,

$$r(\upsilon) = \frac{a(1 - e^2)}{1 + e\cos\upsilon}, \tag{3.74}$$

where a and e are constants of integration—*semimajor axis* and *eccentricity*, respectively:

$$a = \frac{L^2}{\mu GM_1M_2(1 - e^2)}, \qquad e = \left(1 + \frac{2HL^2}{\mu G^2M_1^2M_2^2}\right)^{1/2}. \tag{3.75}$$

Total energy and angular momentum are the constants of motion: energy is constant because there is no external force acting on the system, and angular momentum is constant because the potential is central, i.e., all forces point to the barycenter. Energy and angular momentum are given by the following expressions:

$$H = -\frac{1}{2}\frac{G(M_1 + M_2)\mu}{a}, \qquad L = \mu\omega a^2\sqrt{1 - e^2}. \tag{3.76}$$

Depending on the value of H, orbits can be elliptical[2] ($0 \leqslant e < 1$, $H < 0$), parabolic ($e = 1$, $H = 0$), or hyperbolic ($e > 1$, $H > 0$).

So how does the semimajor axis of this fictitious body's orbit relate to the actual bodies in a two-body system? The center of mass of the system in any coordinate system is given by:

$$\boldsymbol{R} = \frac{M_1\boldsymbol{r}_1 + M_2\boldsymbol{r}_2}{M_1 + M_2}, \tag{3.77}$$

where $\boldsymbol{r}_1$ and $\boldsymbol{r}_2$ are vectors from the origin to the respective bodies. If we denote with $\boldsymbol{r}_1'$ and $\boldsymbol{r}_2'$ vectors from the center of mass to the respective bodies at any particular

[1] The two descriptions are mathematically equivalent.

[2] Circular orbits are a special case of elliptical orbits where the energy is minimized.

instant, we quickly see that the distance vector between the two bodies can be written as:

$$\boldsymbol{r} = \boldsymbol{r}_2 - \boldsymbol{r}_1 = \boldsymbol{r}_2' - \boldsymbol{r}_1'. \tag{3.78}$$

Thus, for each body it follows that:

$$\begin{aligned} \boldsymbol{r}_1' &= \boldsymbol{r}_1 - \boldsymbol{R} = \boldsymbol{r}_1 - \frac{M_1\boldsymbol{r}_1 + M_2\boldsymbol{r}_2}{M_1 + M_2} = -\frac{\mu}{M_1}\boldsymbol{r}, \\ \boldsymbol{r}_2' &= \boldsymbol{r}_1 - \boldsymbol{R} = \boldsymbol{r}_2 - \frac{M_1\boldsymbol{r}_1 + M_2\boldsymbol{r}_2}{M_1 + M_2} = +\frac{\mu}{M_2}\boldsymbol{r}. \end{aligned} \tag{3.79}$$

This implies that each of the two bodies moves in an orbit about the center of mass in the same way that the fictitious body does, except that the size of each orbit is shrunk by μ/M_i and the direction of motion of the two bodies is opposite (due to the difference in signs). As Equation (3.79) holds for any instant in time, it holds when $\boldsymbol{r}_1' \equiv \boldsymbol{a}_1$ and $\boldsymbol{r}_2' \equiv \boldsymbol{a}_2$; from this it follows that the sum of individual semimajor axes gives the semimajor axis of the fictitious body: $a = a_1 + a_2$.

How about eccentricities and angular velocities? Figure 3.9 depicts a motion of both bodies on elliptical orbits. Due to the conservation of angular momentum, the line connecting the centers of both stars contains the center of mass at all times, so eccentricities of both orbits and angular velocities of both bodies must be exactly the same as the eccentricity of the fictitious orbit and the angular velocity of the fictitious body: $e_1 = e_2 = e$ and $\omega_1 = \omega_2 = \omega$.

Finally, what about coplanarity? As angular momentum is conserved, that means that both orbits lie perpendicular to the angular momentum vector, i.e., they are coplanar with respect to one another and with respect to the fictitious body orbit.

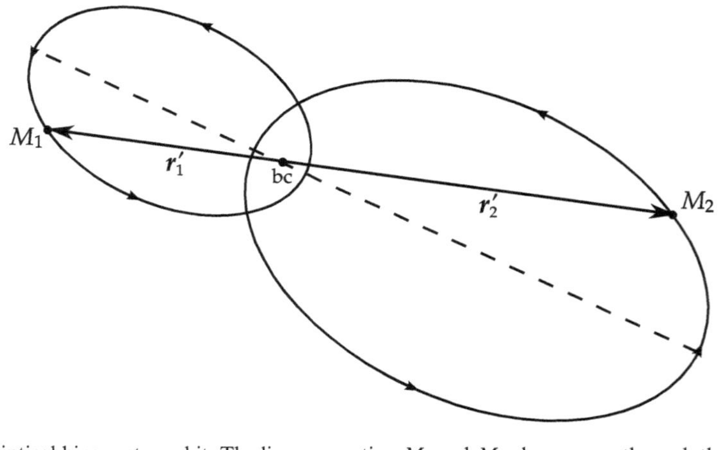

Figure 3.9. Elliptical binary star orbit. The line connecting M_1 and M_2 always goes through the barycenter of the system. In consequence, the orbital period and eccentricity are the same for both orbits.

3.3.1 Orbit Orientation with Respect to Plane of Sky

While the Roche coordinate system lends itself readily for describing the shapes of distorted stars, it is inconvenient for tracking orbital motion of bodies as we observe them from Earth. For that reason we introduce a plane-of-sky coordinate system. It is defined by three axes: u pointing due east, v pointing due north, and w pointing to the observer. The origin of the plane-of-sky coordinate system is in the barycenter at time t_0.

The orientation of an orbit with respect to the observer's plane of sky can be parameterized by three parameters: *inclination* i, *argument of periastron* ω, and *longitude of the ascending node* Ω. Figure 3.10 depicts these three orbital parameters.

INCLINATION i is the angle between the orbital plane and plane of sky. It can assume any value between 0 and π. If $i = 0$, the two planes coincide. For any inclination different from 0, the two planes intersect at two orbital points: an *ascending* node and a *descending* node. In these two nodes the orbiting body approaches and recedes from the observer, respectively. If the inclination is exactly $\pi/2$, the two planes are perpendicular. For $0 \leqslant i < \pi/2$, the motion is *prograde*; for $\pi/2 < i \leqslant \pi$, the motion is *retrograde*.

ARGUMENT OF PERIASTRON ω is the angle from the ascending node to periastron, measured in the direction of motion. It can assume any value between 0 and 2π. By convention, the argument of periastron of the system, ω, equals the argument of periastron of the secondary star orbit, ω_2, and is offset by π from the argument of periastron of the primary star orbit, $\omega = \omega_1 + \pi$.

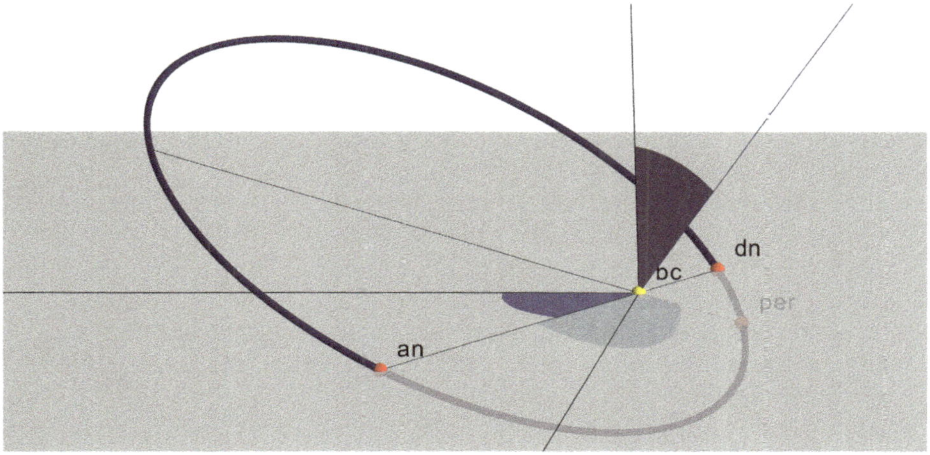

Figure 3.10. Orbital parameters determining the orientation of the orbit: inclination i (purple), argument of periastron ω (cyan), and longitude of the ascending node Ω (blue). The observer is in the direction of the w-axis, and uv is the plane of sky. The motion of the body is counterclockwise, i.e., it crosses the plane of sky in the ascending node while receding from the observer. A Python code for this figure is available online at http://iopscience.iop.org/book/978-0-7503-1287-5.

LONGITUDE OF THE ASCENDING NODE Ω is the angle from the reference direction (typically north, i.e., the v-axis) to the ascending node, measured in a positive (counterclockwise) direction.

If we project the line of sight onto the orbital plane, we obtain two intersections with each of the orbits. These intersections are called *conjunctions*. The intersection that comes after the ascending node is called *superior conjunction*. The intersection that comes after the descending node is called *inferior conjunction*. Conjunctions happen for both components at the same time. These are used to designate binary system components:

PRIMARY STAR is the one that passes through superior conjunction after it goes through the ascending node. It will be *farther* than (behind) the other star at superior conjunction.

SECONDARY STAR is the one that passes through inferior conjunction after it goes through the descending node. It will be *closer* than (in front of) the other star at superior conjunction.

To transform the orientations of the primary and secondary star orbits from the orbital coordinates to plane-of-sky coordinates, we need to perform a series of rotations:

$$\begin{aligned} \boldsymbol{T}_{\text{prim}} &= \boldsymbol{R}_z(\Omega)\boldsymbol{R}_x(-i)\boldsymbol{R}_z(\omega + \pi), \\ \boldsymbol{T}_{\text{sec}} &= \boldsymbol{R}_z(\Omega)\boldsymbol{R}_x(-i)\boldsymbol{R}_z(\omega), \end{aligned} \tag{3.80}$$

where $\boldsymbol{R}_x$ and $\boldsymbol{R}_z$ are active rotation matrices given by:

$$\boldsymbol{R}_x(\theta) = \begin{pmatrix} 1 & 0 & 0 \\ 0 & \cos\theta & -\sin\theta \\ 0 & \sin\theta & \cos\theta \end{pmatrix}, \quad \boldsymbol{R}_z(\theta) = \begin{pmatrix} \cos\theta & -\sin\theta & 0 \\ \sin\theta & \cos\theta & 0 \\ 0 & 0 & 1 \end{pmatrix}. \tag{3.81}$$

There are several points that we should make here, as they are somewhat intricate and require a bit more thought:

- Individual stars pass through the opposite nodes at the same time: as one star crosses the ascending node, the other star crosses the descending node. The arguments of periastra thus differ by π radians. That is why the rotation about z differs by π for both stars.
- The negative sign of i follows from the designation for the primary star: it is by convention the one at superior conjunction, so it needs to be placed behind the secondary star.
- The u and v directions that define the plane-of-sky coordinate system are arbitrarily set. We choose E and N, respectively, but any other choice would work equally well. This in turn implies that the reference direction for the longitude of the ascending node Ω is also set by convention to be 0 due North and increasing in the positive direction (see Figure 3.10).

- Inclinations between π and 2π exchange the roles of the ascending and descending nodes. To avoid confusion, inclination is usually constrained to the $[0, \pi]$ range. In the case where it is not constrained, a *reference node* is used instead of an ascending node to define ω: for inclinations smaller than π, the reference node is the ascending node; for inclinations larger than π, the reference node is the descending node.
- In the case of unresolved binaries, Ω cannot be determined, as there is perfect symmetry with respect to rotations about the line of sight. In the case of astrometric binaries, Ω can be determined, but not uniquely: we cannot distinguish between ascending and descending nodes (i.e., the sign of inclination); thus, we cannot distinguish between $\Omega < \pi$ and $\Omega > \pi$. By convention, purely astrometric solutions provide the value $\Omega < \pi$. In the case of astrometric binaries that are also spectroscopic, radial velocities (RVs) break that degeneracy, and a unique Ω can be obtained.

Let us write out the matrices T_{prim} and T_{sec} given by Equation (3.80) in expanded form, as individual matrix elements will come in handy to qualify eclipses and RVs:

$$T_{\text{prim}} = \begin{pmatrix} -\cos\Omega\cos\omega + \sin\Omega\cos i\sin\omega & \cos\Omega\sin\omega + \sin\Omega\cos i\cos\omega & -\sin\Omega\sin i \\ -\sin\Omega\cos\omega - \cos\Omega\cos i\sin\omega & \sin\Omega\sin\omega - \cos\Omega\cos i\cos\omega & \cos\Omega\sin i \\ \sin i\sin\omega & \sin i\cos\omega & \cos i \end{pmatrix}, \tag{3.82}$$

$$T_{\text{sec}} = \begin{pmatrix} \cos\Omega\cos\omega - \sin\Omega\cos i\sin\omega & -\cos\Omega\sin\omega - \sin\Omega\cos i\cos\omega & -\sin\Omega\sin i \\ \sin\Omega\cos\omega + \cos\Omega\cos i\sin\omega & -\sin\Omega\sin\omega + \cos\Omega\cos i\cos\omega & \cos\Omega\sin i \\ -\sin i\sin\omega & -\sin i\cos\omega & \cos i \end{pmatrix}. \tag{3.83}$$

3.3.2 Stellar Horizon

The boundaries of stars are defined as isosurfaces of the corresponding scalar potential function Ω. If we observe a smooth isosurface at a reference value Ω_0,

$$\Omega(\boldsymbol{r}) = \Omega_0, \qquad \boldsymbol{r} \in \mathbb{R}^3 \tag{3.84}$$

from direction $\hat{\boldsymbol{w}}$, the horizon is a 1D manifold that satisfies the equation:

$$\boldsymbol{g}(\boldsymbol{r}) \cdot \hat{\boldsymbol{w}} = 0, \tag{3.85}$$

where $\boldsymbol{g}(\boldsymbol{r}) = \nabla\Omega$ is the gradient of the potential. The horizon of a smooth surface is a finite set of closed curves, which divides the isosurface into visible and invisible parts with respect to the external observer.

Let $\boldsymbol{r}(s)$ represent a parameterized horizon. Its variation over the surface is described by the following differential equation:

$$\frac{d\boldsymbol{r}}{ds} = \frac{(\boldsymbol{H}(\boldsymbol{r}) \cdot \hat{\boldsymbol{w}}) \times \boldsymbol{g}(\boldsymbol{r})}{\|(\boldsymbol{H}(\boldsymbol{r}) \cdot \hat{\boldsymbol{w}}) \times \boldsymbol{g}(\boldsymbol{r})\|}, \tag{3.86}$$

where $\boldsymbol{H}(\boldsymbol{r}) = (\nabla \otimes \nabla)\Omega(\boldsymbol{r})$ is the Hessian matrix of the potential Ω. The horizon is then obtained by integration of Equation (3.86) from a starting point $\boldsymbol{r}_0$ that is obtained by solving the system of Equations (3.84) and (3.85) iteratively:

$$\boldsymbol{r}_{k+1} = \boldsymbol{r}_k + \delta\boldsymbol{r}(\boldsymbol{r}_k), \tag{3.87}$$

where $\delta\boldsymbol{r}$ is the iteration step,

$$\delta\boldsymbol{r}(\boldsymbol{r}) = \alpha\, \boldsymbol{g}(\boldsymbol{r}) + \beta\, \boldsymbol{H}(\boldsymbol{r}) \cdot \hat{\boldsymbol{w}}, \tag{3.88}$$

with the weights α and β calculated via:

$$\begin{bmatrix} \alpha \\ \beta \end{bmatrix} = \begin{bmatrix} \|\boldsymbol{g}(\boldsymbol{r})\|^2 & \boldsymbol{g}(\boldsymbol{r}) \cdot \boldsymbol{H}(\boldsymbol{r}) \cdot \hat{\boldsymbol{w}} \\ \boldsymbol{g}(\boldsymbol{r}) \cdot \boldsymbol{H}(\boldsymbol{r}) \cdot \hat{\boldsymbol{w}} & \|\boldsymbol{H}(\boldsymbol{r}) \cdot \hat{\boldsymbol{w}}\|^2 \end{bmatrix}^{-1} \begin{bmatrix} \Omega_0 - \Omega(\boldsymbol{r}) \\ -\boldsymbol{g}(\boldsymbol{r}) \cdot \hat{\boldsymbol{w}} \end{bmatrix}. \tag{3.89}$$

The iteration is initiated from any point $\boldsymbol{r}_0$ that lies inside the isosurface.

3.3.3 Apsidal Motion

This would be a good time to recall our digression in Section 3.2 where we pointed out that tidal interactions and rotation cause a redistribution of mass in stellar interiors. As noted then, this cannot be modeled appropriately without the density distribution model, but even in the absence of such a model, it is clear that this bulging causes torque in the system and consequently *rotation* of the line of apsides. This rotation is referred to as *apsidal motion.*

Apsidal motion is very common in close binaries and should thus be carefully handled: if ω changes, then the effective period of the binary no longer corresponds to the true sidereal period, but to the anomalous period (the time between two consecutive periastron passages). In addition, the phasing of the eclipses will also change. That is why special care should be exercised when defining orbital phase. That is the topic of our next section.

3.4 Dynamical Aspects of Binary Systems

The formulation of dynamical aspects of binary stars differs from the historical (Russell's) formulation given, e.g., by Smart & Green (1977); textbooks and papers that are paving the field of binary stars, i.e., Wilson & Devinney (1971), Etzel (1981), Hadrava (1990), Hilditch (2001), Kallrath & Milone (2009), and others, typically do not discuss this formulation thoroughly; that is why we focus on it here in some detail.

3.4.1 Ephemeris Parameters

When describing dynamical aspects of binary systems, we typically use a set of quantities that encapsulate the periodicity of observations. This set of quantities is referred to as the *ephemeris* (from Latin, meaning *journal, diary*; plural *ephemerides*).

There are three quantities that determine the ephemeris of the observed binary:

REFERENCE TIME t_0 is a unique point in time that sets the origin of the ephemeris. Its choice is arbitrary. Practical choices include the time of periastron passage (t_{per}) and the time of superior conjunction (t_{sc}). Reference time is given in days, either as a heliocentric or a barycentric Julian date (see Section 2.1).

ORBITAL PERIOD P_0 is the sidereal time of revolution of both components in a binary system around the barycenter. If the orbital period is changing, the value P_0 corresponds to the value at reference time t_0. It is given in days.

RATE OF PERIOD CHANGE $\dot{P}_0 \equiv dP/dt$ is the first time derivative of the orbital period:

$$\dot{P}_0 = \frac{P(t) - P_0}{t - t_0} \Rightarrow \quad P(t) = P_0 + \dot{P}_0(t - t_0). \tag{3.90}$$

The rate of period change is unitless, given in days day^{-1}.

A typical choice for eclipsing binary stars is to set t_0 to the time of first superior conjunction (deeper minimum) in observed light curves, t_{sc}. The ephemeris is then denoted as:

$$\text{min. I} = t_0(uncert.) + \mathrm{E} \times P_0(uncert.) + \frac{1}{2}\mathrm{E}^2 \times \dot{P}_0(uncert.), \tag{3.91}$$

where E is an integer that enumerates successive superior conjunctions. The uncertainty is typically given in terms of significant figures; for example, 5.603(12) means 5.603 ± 0.012. Equation (3.91) assumes small deviations where $P(t) \approx P_0$ and $\dot{P}_0 \approx$ const. If this is not fulfilled, higher powers of E will need to be included: E^3 for $\ddot{P}_0$, E^4 for $\dddot{P}_0$, etc. If $\dot{P}_0 = 0$, the equation describes a perfectly periodic (cyclical) orbit.

3.4.2 The Kepler Problem

The established approach to solving the *Kepler* problem is by introducing *anomalies*. Figure 3.11 shows an elliptical orbit given by Equation (3.74), with three anomalies depicted:

TRUE ANOMALY υ is the angle measured from periastron to the position of the star on its orbit relative to the focal point of the ellipse. It represents an independent variable in Equation (3.74) that determines the orbit, $r(\upsilon)$. As its rate of change depends on time and this variation cannot be expressed analytically in terms of time, we use an auxiliary circle depicted in Figure 3.11.

ECCENTRIC ANOMALY E is the angle from periastron to the projected point Q on the auxiliary circle relative to the center of the ellipse. The eccentric anomaly can be related to the true anomaly through a simple trigonometric relationship that we now derive.

Let us denote the coordinates of point P in the coordinate system with the origin in O with x_P and y_P; then we can write:

$$x_P = ae + r\cos\upsilon, \quad y_P = r\sin\upsilon. \tag{3.92}$$

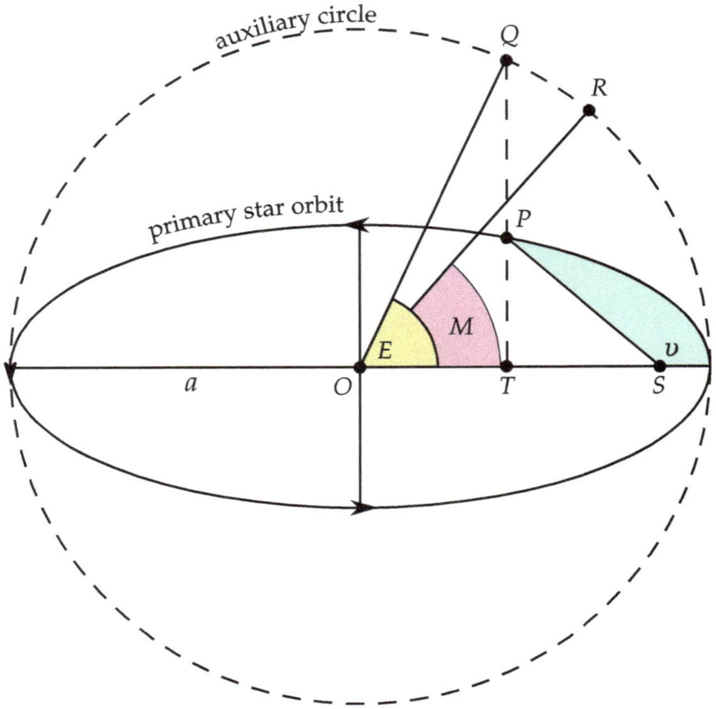

Figure 3.11. Graphical representation of the true anomaly υ, eccentric anomaly E, and mean anomaly M. The star is on point P at the orbit, rotating in a counterclockwise direction around the focal point S. Point Q is constructed by vertical projection of the star's position on the auxiliary circle. The mean anomaly M, equal to the angle ROS, is proportional to the time elapsed since the star's passage through periastron.

Similarly, we can express the coordinates of point Q in the same coordinate system as:

$$x_Q = a \cos E, \quad y_Q = a \sin E. \tag{3.93}$$

Distance $|TS|$ can then be written as:

$$|TS| = ae - a \cos E = r \cos(\pi - \upsilon) = -r \cos \upsilon. \tag{3.94}$$

Expressing $\cos \upsilon$ and plugging it into Equation (3.74) yields the equation of orbit in terms of eccentric anomaly:

$$r = a(1 - e \cos E). \tag{3.95}$$

Coordinates x_P, y_P, x_Q, and y_Q are related to each other through simple identities:

$$\begin{aligned} x_P = x_Q: &\quad ae + a(1 - e \cos E)\cos \upsilon = a \cos E, \\ ay_p = by_q: &\quad ar \sin \upsilon = ba \sin E. \end{aligned} \tag{3.96}$$

From these two equations it follows that:

$$\cos E = \frac{e + \cos \upsilon}{1 + e \cos \upsilon}, \quad \sin E = \frac{\sqrt{1 - e^2}}{1 + e \cos \upsilon} \sin \upsilon. \tag{3.97}$$

Finally, we divide these two equations to obtain a relationship between the true and eccentric anomaly:

$$\tan E = \frac{\sqrt{1 - e^2} \sin \upsilon}{e + \cos \upsilon}. \tag{3.98}$$

This equation can be converted to a more practical, easily invertible form by using half-angle identities (Smart & Green 1977):

$$\tan \frac{E}{2} = \sqrt{\frac{1 - e}{1 + e}} \tan \frac{\upsilon}{2}. \tag{3.99}$$

MEAN ANOMALY M introduces the time dependence explicitly. We define it as:

$$M = \omega(t - t_{\mathrm{per}}) \equiv 2\pi \frac{t - t_{\mathrm{per}}}{P_0}, \tag{3.100}$$

where t_{per} is the time of periastron passage and P_0 is the orbital period of the binary.

The mean anomaly is a measure of time and, as such, cannot be geometrically constructed from the true and/or eccentric anomaly in Figure 3.11, but we can derive the relationship between the mean and eccentric anomaly algebraically. We start with the reciprocal of the equation of orbit:

$$\frac{1}{r} = \frac{1 + e \cos \upsilon}{a(1 - e^2)}, \tag{3.101}$$

and we evaluate its time derivative:

$$-\frac{\dot{r}}{r^2} = -\frac{e \sin \upsilon \dot{\upsilon}}{a(1 - e^2)}. \tag{3.102}$$

This is convenient because we can write out specific angular momentum, $l = r^2 \upsilon'$, given by Equation (3.76):

$$r^2 \dot{\upsilon} = \frac{a(1 - e^2)\dot{r}}{e \sin \upsilon} = \omega a^2 \sqrt{1 - e^2}, \tag{3.103}$$

where $\omega = 2\pi / P_0$. Now we need to find $\dot{r}$ from Equation (3.95):

$$\dot{r} = ae \sin E \dot{E}. \tag{3.104}$$

Expressing $\dot{E}$ and using the y-component alignment from Equation (3.96) yields:

$$\dot{E} = \frac{a\omega \sin \upsilon}{a \sin E \sqrt{1 - e^2}} = \frac{a\omega}{r} = \frac{a\omega}{a(1 - e \cos E)}. \tag{3.105}$$

Finally, we integrate,

$$\int (1 - e \cos E)dE = \int \omega dt \ \Rightarrow \ E - e \sin E = \omega t + C, \tag{3.106}$$

where the integration constant C is determined from the condition $E|_{t=t_{\mathrm{per}}} = 0$, yielding $C = -\omega t_{\mathrm{per}}$:

$$E - e \sin E = \omega(t - t_{\mathrm{per}}) \equiv M. \tag{3.107}$$

This expression cannot be solved for E explicitly, so it needs to be solved iteratively. As a side note we should mention that this is the very reason why solving $r(t)$ analytically is not possible for general Keplerian orbits.

Thus, to obtain $r(t)$, we need to do the following: for each time t we compute M using Equation (3.100), from M we iteratively compute E using Equation (3.107), from E we compute υ using Equation (3.99), and from υ we compute $r(t)$ using Equation (3.74). When true anomaly is of no explicit interest, the obvious shortcut exists where we can compute $r(E)$ directly from Equation (3.95).

3.4.3 Orbital Phase

The orbit, along with any orbital evolution, is fully described by the ephemeris discussed in Section 3.4.1. However, thanks to the periodic or quasi-periodic behavior of eclipsing binary light curves, it is convenient to introduce the *orbital phase* of the binary and use it instead of time.

ORBITAL PHASE Φ is a fraction of the orbital period that has elapsed since the last completed revolution cycle. It is defined on the [0, 1] interval. It is a linear function of time only for circular orbits; for eccentric orbits it is a nonlinear function of time and a linear function of mean anomaly:

$$M = 2\pi\,(\Phi - \Phi_{\mathrm{per}}). \tag{3.108}$$

Orbital phase thus generalizes our definition of epoch E in Equation (3.91): while E is an integer counter that provides the time of superior conjunction, Φ is a real-valued parameter that provides the time of any point in orbit.

Zero phase is defined by orbital properties. The old convention for eclipsing binaries (the so-called Russell convention) used superior conjunction (the mid-time of the deeper eclipse). This convention would be both suitable and convenient (because it is easily measured) if it were not for apsidal motion. Apsidal motion causes excursions of *both* eclipses in opposite directions with respect to a uniformly advancing time. In other words, a binary star with a constant period and apsidal motion would show an anomalous, cyclical period change. That is why we need a more robust definition of zero phase, one that depends not on any particular point on the orbit but on the orbit itself. Thus, zero phase is defined by the argument of periastron and eccentricity in a way that satisfies Equation (3.108). In other words,

$\phi = 0$ corresponds to an *arbitrary* point on the ellipse that yields Φ_{per} consistent with mean anomaly (see Figure 3.12). It *happens* to coincide with the phase of superior conjunction when $\omega = \pi/2$, that is, when the line of sight is aligned with the semimajor axis.

Orbital phase may be visualized much like mean anomaly: by an angle on the auxiliary circle that increases steadily with time from the zero point.

The phase of conjunction is computed by using anomalies. Figure 3.12 shows three important points: the zero phase $\Phi = 0$, the conjunction phase $\Phi = \Phi_C$, and the periastron phase $\Phi = \Phi_{per}$. The first step is to write the true anomaly of superior conjunction, which we infer from Figure 3.12:

$$\upsilon_C = -\left(\omega - \frac{\pi}{2}\right) = \frac{\pi}{2} - \omega. \tag{3.109}$$

Once the true anomaly of the conjunction is known, Equations (3.99) and (3.107) may be used to obtain the mean anomaly at conjunction, M_C. Inspecting Figure 3.12 again, it is immediately evident that:

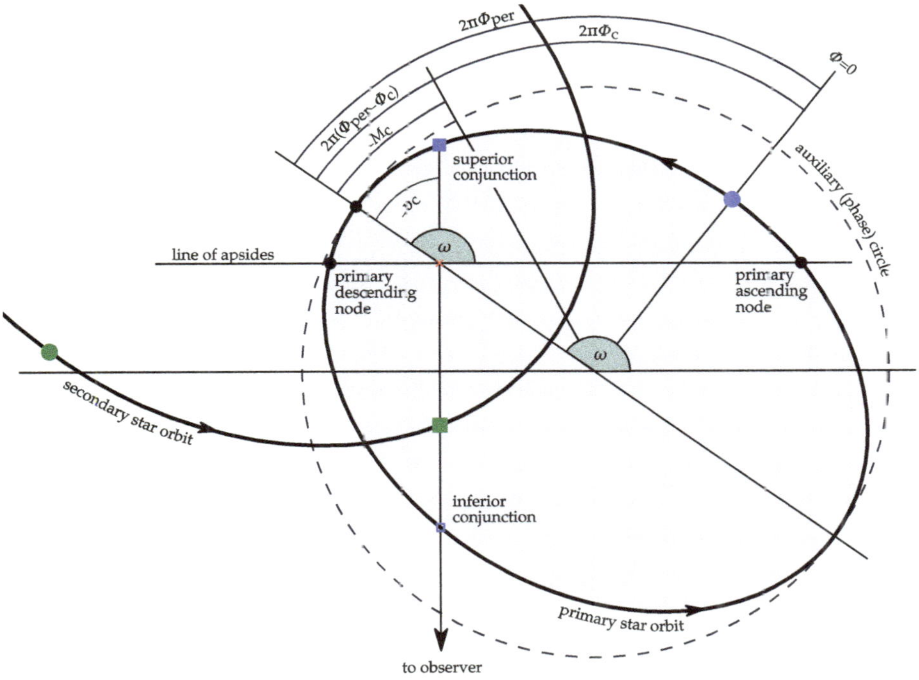

Figure 3.12. Schematic view of the orbital plane. The direction to the observer is downward. The intersection between the orbital plane and the plane of sky yields the line of apsides. The ascending node of the primary star (i.e., primary ascending node) is by definition the one where the object goes away from the observer (into the plane of sky). Zero phase ($\Phi = 0$; blue circle) is defined by the reference time rather than any particular point on the orbit, and it depends on eccentricity e and argument of periastron ω. Superior and inferior conjunctions are denoted with a filled and open square, respectively. Phases of superior conjunction (Φ_c) and periastron (Φ_{per}) are computed by Equations (3.109)–(3.111).

$$2\pi(\Phi_{\text{per}} - \Phi_C) = -M_C \quad \text{and} \quad 2\pi\Phi_{\text{per}} = \omega - \frac{\pi}{2}. \tag{3.110}$$

The phase of conjunction may then be computed by the following expression:

$$\Phi_C = \frac{M_C + \omega}{2\pi} - \frac{1}{4}. \tag{3.111}$$

To demonstrate this formalism on a particular example, consider a detached eclipsing binary with $e = 0.3$ and $\omega = \pi/4$. From Equation (3.109) we see that the true anomaly of conjunction $\upsilon_C = \pi/2 - \pi/4 = \pi/4$. Using Equations (3.99) and (3.107), we obtain $E_C = 0.59$ and $M_C = 0.42$. Using Equations (3.110)–(3.111), we get the phase of periastron $\Phi_{\text{per}} = -0.125$ and the phase of (superior) conjunction $\Phi_C = -0.06$.

PHASE SHIFT $\Delta\Phi$ is a constant offset to the orbital phase, introduced as a matter of convenience:

$$\Phi = \left(\frac{t - t_0}{P(t)} \bmod 1\right) + \Delta\Phi. \tag{3.112}$$

The value of $\Delta\Phi$ is usually 0, but there are several cases when its value is different from 0:

1. Some still prefer to set the value of t_0 to correspond to the superior conjunction because of the very efficient methods for determining the times of minima (e.g., Kwee & van Woerden 1956); $\Delta\Phi$ should then be set to the value of Φ_C.
2. Available data may already have been transformed to phases, and the original time stamps of observations are not available; $\Delta\Phi$ can then be used as a free parameter.
3. The roles of both stars may need to be exchanged: a suitable $\Delta\Phi$ should then be applied to the ephemeris.

To compute phase as a function of time, we need to solve the following differential equation:

$$d\Phi = \frac{dt}{P(t)} \bmod 1. \tag{3.113}$$

If $P \neq P(t)$, solving this by simple integration yields (without the 2π factor) Equation (3.100). If, on the other hand, P changes linearly in time, we have:

$$\begin{aligned} \Phi(t) &= \int \frac{dt}{P_0 + \dot{P}_0(t - t_0)} \bmod 1 \\ &= \frac{1}{\dot{P}_0} \log[P_0 + \dot{P}(t - t_0)] \bmod 1. \end{aligned} \tag{3.114}$$

The direct consequence of this time-to-phase transformation is that phase points will be smeared if apsidal motion is present in the system. Since apsidal motion changes argument of periastron ω, conjunction phase changes according to Equations (3.110)–(3.111). If we defined zero phase to coincide with conjunctions, we would be measuring an *anomalous* orbital period rather than the true orbital period, and the changes in the light-curve shape due to the changing value of ω would cause smearing of the secondary eclipse in the phased light curve.

3.4.4 Volume Conservation in Eccentric Orbits

A nontrivial assumption that partly follows from physical insight and partly from observations is stellar volume conservation in eccentric orbits. One of the fundamental assumptions of the Roche model is that a star adapts to the instantaneous force field on a timescale that is significantly shorter than the orbital period. In other words, a star will instantly change its shape to adapt to the equipotential that bounds it. In the case of eccentric orbits, however, equipotential lobes can change their volumes substantially. If a star adapted to this instantaneous equipotential at every point in its orbit, that would mean that it needs to rarefy or compress in order to fill the volume of the equipotential. In that case, some of the heat generated during compression would be irradiated away and thus not recovered during subsequent expansion. This energy would need to be replaced from the energy stored in the orbit, leading to rapid circularization since even a tiny energy loss compounds quickly over many orbits. In effect, there would be essentially no eccentric short-period binaries in the sky. This would also significantly impact the efficiency of Kozai cycles (Kozai 1962; Mazeh & Shaham 1979), which tighten the orbits on timescales orders of magnitude longer than the orbital period.

Observationally, if the volumes of stars changed on eccentric orbits, we would expect to see this effect in the measured radii of eclipsing binaries as a function of the argument of periastron. However, eclipse durations are consistent with near-constant radii and, thus, near-constant volumes. Even the red giant and supergiant stars that would arguably be most susceptible to volume change do not exhibit an appreciable change in radius as a function of orbital phase. The star does not adapt to the equipotential, but instead conserves its volume and fills a *different* equipotential from that at periastron. We thus conclude that volume conservation is a safe assumption with strong footing in both theoretical insight and observations.

That said, there is a particularly well-suited test bed for studying the validity of volume conservation in the limit of high eccentricities: highly eccentric ellipsoidal variables (also known as *heartbeat* stars; Thompson et al. 2012) that host red giants. Their components at periastron are separated by only a few stellar radii, giving rise to significant tidal forces that distort stellar surfaces. A total of 28 giants in heartbeat stars have been reported to date (Nicholls & Wood 2012; Gaulme et al. 2013, 2013; Beck et al. 2014; Richardson et al. 2016); of these, there are nine for which the authors developed a full model (Nicholls & Wood 2012; Beck et al. 2014; Richardson et al. 2016), all of them under the constant-volume assumption. Yet these are the prime candidates where small deviations from constant volume are plausible.

3.4.5 Light Curves

The total light emitted from a binary star in the direction of the observer is the net sum of light coming from all surface elements over the visible surface. We defer the full discussion of the radiative properties of binary stars until Chapter 4; here we only stress that light variation across a stellar surface depends on many factors. These include the distribution of effective temperature and surface gravity across the tidally and rotationally distorted surfaces of the stars, reflection and heating due to mutual irradiation, limb darkening, spots, pulsations, eclipses, etc. Each local surface element contributes to the total light according to the net surface brightness distribution. As stars move in their orbits and rotate about their axes, the contribution of each local surface element changes as the angles with respect to vantage point change. Consequently, the net amount of light the observer receives from the system depends on time. Plotting light as a function of time or as a function of phase is called a *light curve*. Figure 3.13 depicts a light curve for the eccentric binary example from Section 3.4.3.

3.4.6 Radial Velocities

In Section 3.3.1 we derived a transformation from the orbital plane to the plane of sky. The derivative of the w-component (along the line of sight) defines RV; for the primary star,

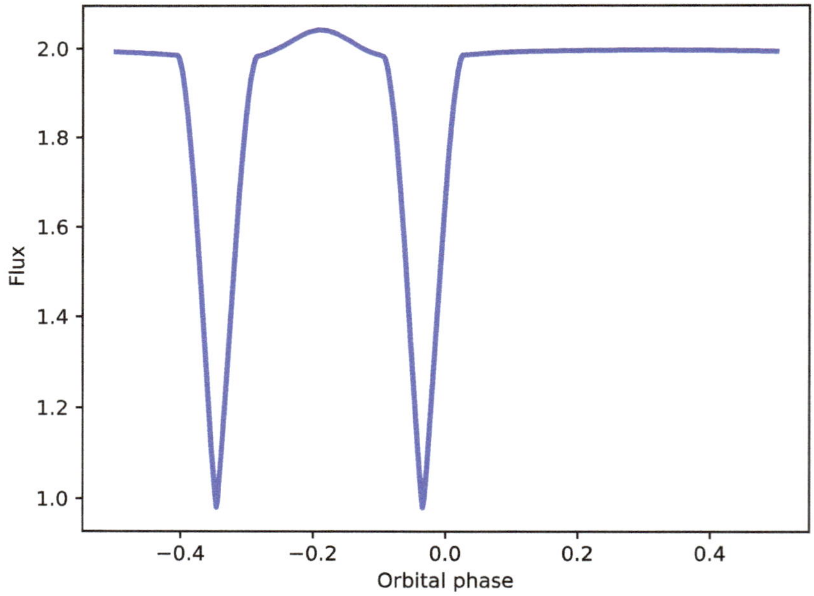

Figure 3.13. Light curve of the eccentric binary system discussed in Section 3.4.3: $e = 0.3$, $\omega = \pi/4$ result in the phase of (superior) conjunction at $\Phi_C = -0.06$.

$$\begin{pmatrix} \dot{u} \\ \dot{v} \\ \dot{w} \end{pmatrix} = \boldsymbol{T}_{\text{prim}} \begin{pmatrix} \dot{x} \\ \dot{y} \\ 0 \end{pmatrix}. \tag{3.115}$$

To calculate $\dot{w}$, we need $\dot{x}$ and $\dot{y}$; these are readily obtained from Equation (3.96):

$$\begin{aligned} x &= a_1 \cos E \;\Rightarrow\; \dot{x} = -a_1 \dot{E} \sin E, \\ y &= b_1 \sin E \;\Rightarrow\; \dot{y} = b_1 \dot{E} \cos E, \end{aligned} \tag{3.116}$$

where a_1 and b_1 denote the semimajor and semiminor axes of the primary star's orbit. The eccentric anomaly derivative $\dot{E}$ follows from Kepler's equation, Equation (3.107):

$$\dot{M} = \omega = \dot{E}(1 - e \cos E) \;\Rightarrow\; \dot{E} = \frac{\omega}{1 - e \cos E}. \tag{3.117}$$

Thus,

$$\dot{x} = -\frac{a_1 \omega \sin E}{1 - e \cos E}, \quad \dot{y} = \frac{b_1 \omega \cos E}{1 - e \cos E}, \tag{3.118}$$

and the RV is:

$$\begin{aligned} \dot{w} &= \sin i \sin \omega \dot{x} + \sin i \cos \omega \dot{y} \\ &= \frac{a_1 \omega \sin i}{1 - e \cos E}\left(\sqrt{1 - e^2} \cos \omega \cos E - \sin \omega \sin E\right). \end{aligned} \tag{3.119}$$

This expression can be converted into a more convenient form by substituting $\sin E$ and $\cos E$ with relations given in Equation (3.97):

$$\begin{aligned} \text{RV}_1 \equiv \dot{w} &= \frac{2\pi a_1 \sin i}{P\sqrt{1 - e^2}}[e \cos \omega + \cos(\omega + \upsilon)] \\ &= \frac{2\pi q a \sin i}{P(1 + q)\sqrt{1 - e^2}}[e \cos \omega + \cos(\omega + \upsilon)]. \end{aligned} \tag{3.120}$$

Factor $K_1 \equiv 2\pi q a \sin i / P(1 + q)\sqrt{1 - e^2}$ is the semi-amplitude of the primary star's RV curve. Similarly, for the secondary star,

$$\text{RV}_2 = -\frac{2\pi a \sin i}{P(1 + q)\sqrt{1 - e^2}}[e \cos \omega + \cos(\omega + \upsilon)]. \tag{3.121}$$

with the corresponding semi-amplitude $K_2 \equiv 2\pi a \sin i / P(1 + q)\sqrt{1 - e^2}$. Equations (3.120) and (3.121) describe the motion with respect to the barycenter of the system; if there is any radial component of the barycentric (systemic, center-of-mass) velocity, typically denoted with v_γ, it needs to be added to both equations. Figure 3.14 depicts an *RV curve*: RV as a function phase for each component in a binary system.

RVs of extended bodies are, much like their fluxes, the net sum of RVs of all visible surface elements, weighted by their contribution to total light in the direction

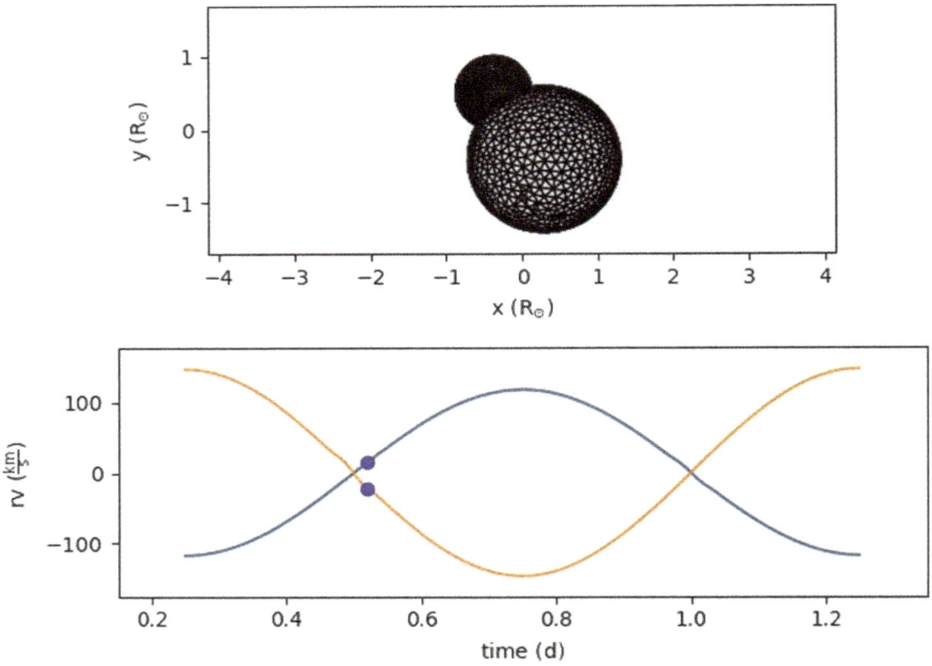

Figure 3.14. RV curve of a binary system with mass ratio $q = 0.8$, inclination $i = 80°$, and orbital period $P_0 = 1$ day. The figure in the interactive version of the book is animated, available online at https://doi.org/10.1088/978-0-7503-1287-5.

of the observer. An extended discussion on the effects of tidal distortion on RV curves is given in Wilson & Sofia (1976).

3.4.7 Eclipsing Systems

The crucial advantage of eclipsing over noneclipsing binary systems is the presence of eclipses, which are used as indicators of sizes and surface structure of individual components. Depending on orbital inclination, eclipsing binaries may exhibit the following:

PARTIAL ECLIPSES, where the stars partially overlap. The shape of eclipses is typically sharp, V-shaped. Figure 3.15 depicts an example of a partially eclipsing binary: its light curve, shape, and orientation. The interactive version of the book features an animated plot.

TOTAL ECLIPSES, where the stars fully overlap. The profile of eclipses is typically box-shaped, since the eclipses are either total or annular. Figure 3.16 depicts an example of a totally eclipsing binary: its light curve, shape, and orientation. Note that the primary (annular) eclipse is *not* flat: this is due to a nonuniform brightness distribution over the primary star disk.

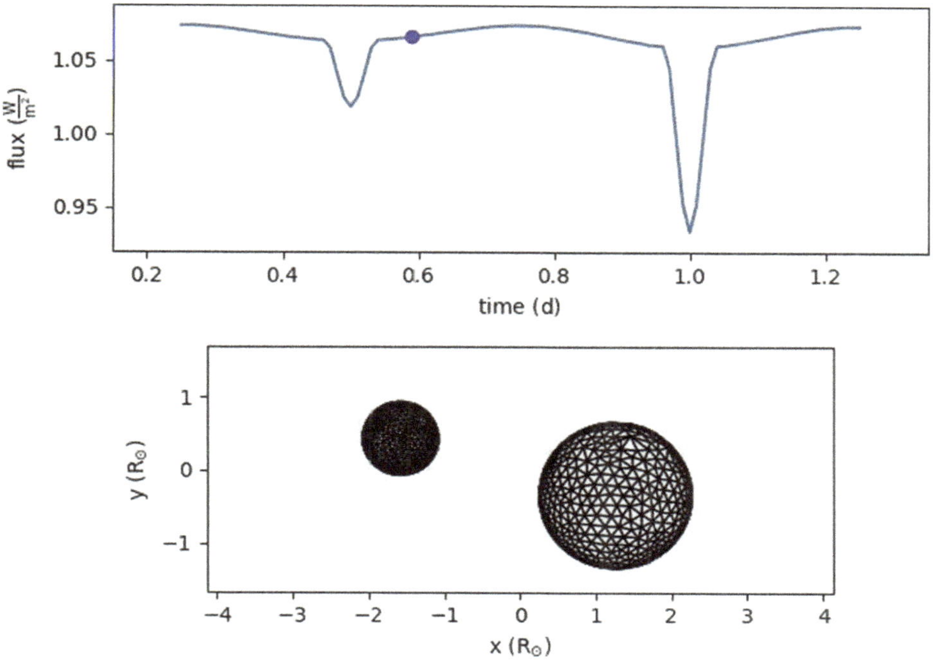

Figure 3.15. Partially eclipsing binary system. The ratio of the radii is 0.5, the ratio of the temperatures is 0.8, and the inclination is 80°. The interactive version of this figure is animated, available online at https://doi.org/10.1088/978-0-7503-1287-5.

The condition for the eclipse on circular orbits is readily determined by expressing the position of the eclipsing star in the plane-of-sky coordinates (see Equation (3.83)):

$$\begin{aligned} u &= a[\cos\Omega\cos(\omega+\upsilon) - \sin\Omega\cos i\sin(\omega+\upsilon)], \\ v &= a[\sin\Omega\cos(\omega+\upsilon) + \cos\Omega\cos i\sin(\omega+\upsilon)], \\ w &= -a\sin i\sin(\omega+\upsilon). \end{aligned} \quad (3.122)$$

The projection onto the plane of sky uv is then:

$$\Delta^2 = u^2 + v^2 = a^2[\cos^2(\omega+\upsilon) + \sin^2(\omega+\upsilon)\cos^2 i]. \quad (3.123)$$

The condition for the eclipse can then be written as:

$$\begin{aligned} R_1 + R_2 < \Delta \quad &: \text{no eclipses} \\ R_1 - R_2 < \Delta < R_1 + R_2 \quad &: \text{partial eclipse} \\ \Delta < R_1 - R_2 \quad &: \text{total eclipse.} \end{aligned} \quad (3.124)$$

An analogous criterion can be used for eccentric orbits if a in Equation (3.123) is replaced by the instantaneous separation $r(\upsilon)$ from Equation (3.74).

Eclipses also affect RV curves. In the previous section we stressed that the net RVs of extended bodies are a sum of the visible elements' RVs, weighted by the

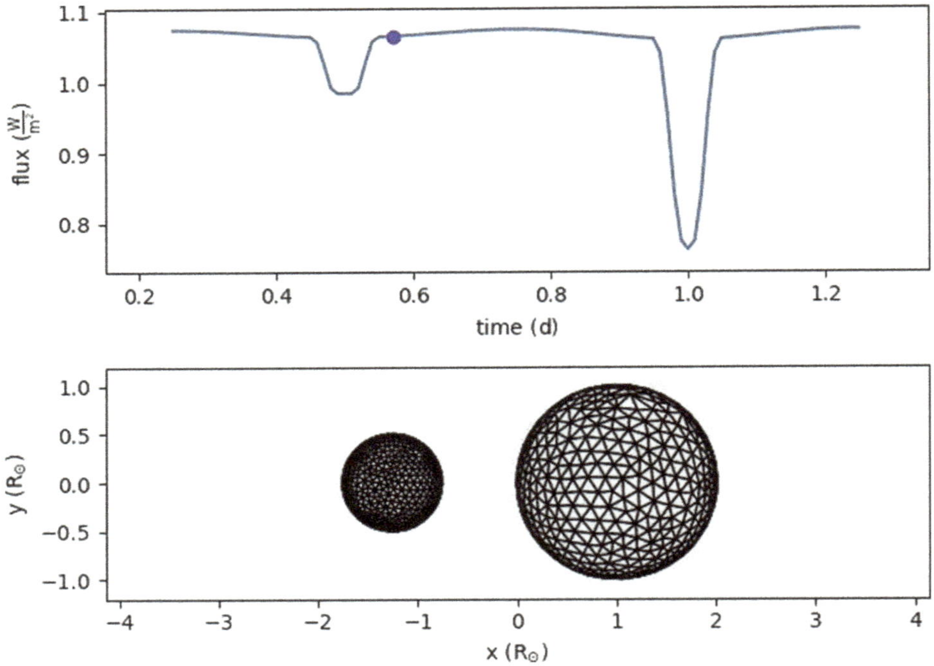

Figure 3.16. Totally eclipsing binary system. The ratio of the radii is 0.5, the ratio of the temperatures is 0.8, and the inclination is 90°. The interactive version of this figure is animated, available online at https://doi.org/10.1088/978-0-7503-1287-5.

amount of light originating from those elements in the direction of the observer. Eclipses change the visibility of elements and, consequently, the net RV. This is exhibited by the "kink" in RV curves, demonstrated in Figure 3.17, and known as the Rossiter–McLaughlin effect after Rossiter (1924) and McLaughlin (1924).

The Rossiter–McLaughlin effect is symmetrical about the dynamical RV curve when the spin axes are aligned with the orbital axis, and asymmetrical in the case of misaligned axes. We will return to this in Section 6.1.

3.4.8 Ellipsoidal Variations

Another significant effect arises as a result of stellar surface distortion, which is aspect dependent: variation of the overall flux caused by the variation of the visible cross section of the star. As the star is oblate owing to tidal forces and rotation, we see a larger surface cross section when the stars are oriented sideways, i.e., when their plane-of-sky separation is the largest. Conversely, we see a smaller surface cross section when we are looking along the semimajor axis. This effect is called ellipsoidal variation of the light curve; Figure 3.18 shows an example of ellipsoidal variation of a noneclipsing semi-detached binary with depicted orientation of both components in critical phases. The effect is present even if a binary is not eclipsing, for as long as the inclinations are not too low, as there is no variation in the cross section for face-on orbits.

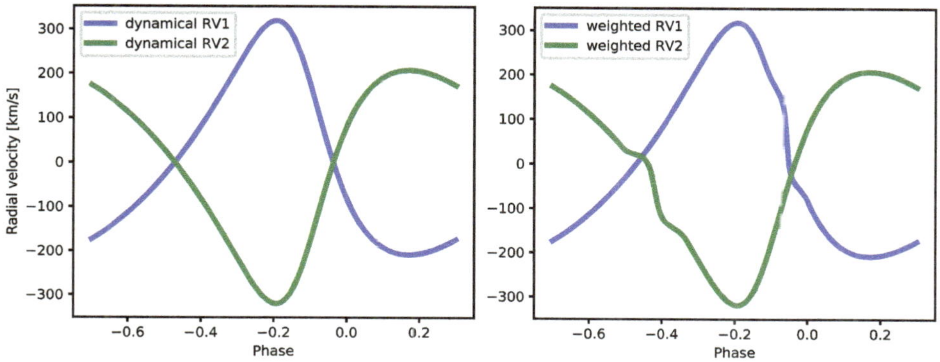

Figure 3.17. Rossiter–McLaughlin effect for the same eclipsing binary as in Figure 3.13. The left panel depicts a dynamical RV curve, i.e., the RV curve of the center of mass of the star. The right panel depicts a mesh RV curve, computed as a weighted average of RVs of all visible surface elements.

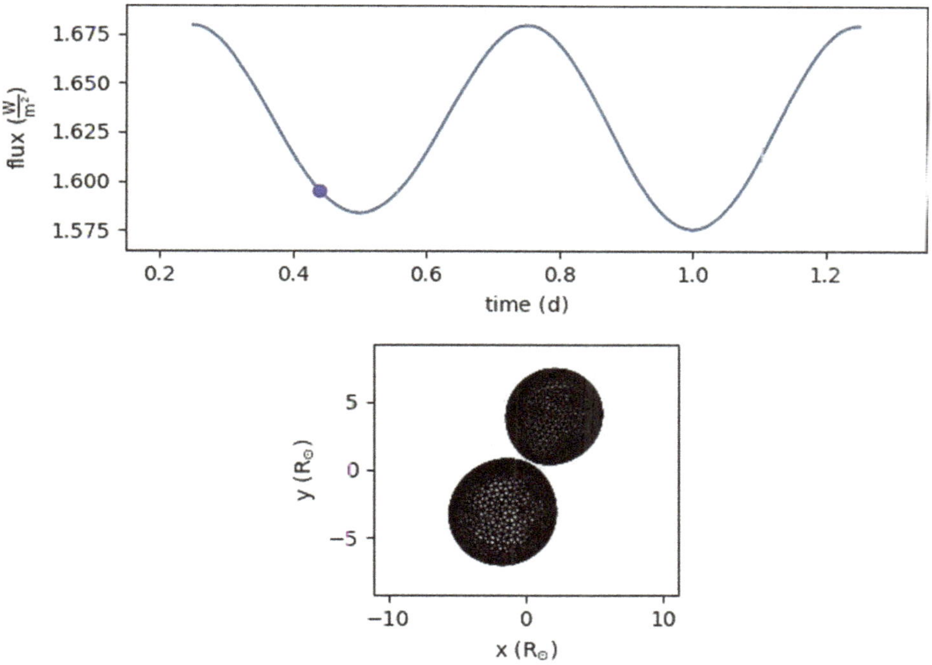

Figure 3.18. Light-curve variation of a close detached binary due to ellipsoidal variations. Shape distortion causes the cross sections of the stars visible to the observer to change with phase. The inclination of the system $i = 40°$ is chosen so that the light curve exhibits no eclipses: the variation is exclusively due to ellipsoidal variations. The figure in the interactive version of the book is animated, available online at https://doi.org/10.1088/978-0-7503-1287-5.

3.4.9 Eccentricity and Argument of Periastron from Light-curve Geometry

The phases at which eclipses occur depend on the orbital properties of the binary star. The separation $\Delta\Phi$ between eclipses is proportional to the tangential component of eccentricity, $e \cos \omega$, and the ratio of eclipse durations, $(d_s - d_p)/(d_s + d_p)$, is proportional to the radial component of eccentricity, $e \sin \omega$. For small eccentricities and inclinations close to 90°, we can write (Paddock 1913):

$$e \cos \omega = \sqrt{1 - e^2} \tan\left(\pi\Delta\Phi - \frac{\pi}{2}\right), \quad e \sin \omega = \frac{d_s - d_p}{d_s + d_p}. \tag{3.125}$$

For large eccentricities, an iterative Newton–Raphson approach is necessary; if we define an auxiliary quantity ψ (Kopal 1959, Equation (9-8)),

$$\psi = \pi + 2 \arctan \frac{e \cos \omega}{\sqrt{1 - e^2}}, \tag{3.126}$$

then Kepler's second law may be written as an iterative equation:

$$2\pi\Delta\Phi = \psi - \sin \psi. \tag{3.127}$$

Solving Equation (3.127) for ψ, expressing $e \cos \omega$ from Equation (3.126), and combining it with the expression for $e \sin \omega$ from Equation (3.125), we obtain:

$$e = \left[\sin^2\left(\frac{\psi - \pi}{2}\right) + \left(\frac{d_s - d_p}{d_s + d_p}\right)^2 \cos^2\left(\frac{\psi - \pi}{2}\right)\right]^{1/2}. \tag{3.128}$$

Conversely, we can compute the argument of periastron from either:

$$\begin{aligned} \omega &= \arcsin\left(\frac{1}{e} \frac{d_s - d_p}{d_s + d_p}\right) \text{ or} \\ \omega &= \arccos\left(\frac{1}{e}\sqrt{1 - e^2} \tan \frac{\psi - \pi}{2}\right), \end{aligned} \tag{3.129}$$

with solution multiplicity determining the quadrant of ω.

In general, $e \cos \omega$ is attainable to a notably higher precision than $e \sin \omega$; because of that, systematics may often arise in an attempt to derive e and ω from eclipse times and Equations (3.128)–(3.129) that prefers the values $\omega \approx \pm\pi/2$ and small but nonzero eccentricities. To mitigate these systematics, a reasonable estimate for e may be obtained by taking the average value for $\sin^2 \omega = 1/2$ and rewriting Equation (3.128):

$$\begin{aligned} e &= \sin \frac{\psi - \pi}{2}\left[1 - \sin^2 \omega \cos^2 \frac{\psi - \pi}{2}\right]^{-1/2} \\ &= \sin \frac{\psi - \pi}{2}\left[1 - \frac{1}{2} \cos^2 \frac{\psi - \pi}{2}\right]^{-1/2}. \end{aligned} \tag{3.130}$$

There is another important point to be realized when dealing with eccentric orbits: even if the effective temperatures of both components are the same, the depths of the minima are different. This is due to the fact that the fraction of the eclipsed stellar surface depends on the distance between the stars on orbit. It is also possible, due to this point, to have only one eclipse, when the stars are near periastron, and that the other eclipse is completely missing.

References

Beck, P. G., Hambleton, K., Vos, J., et al. 2014, A&A, 564, A36
Chandrasekhar, S. 1933, MNRAS, 93, 462
Ester, M., Kriegel, H., Sander, J., & Xu, X. 1996, in Proc. 2nd Int. Conf. on Knowledge Discovery and Data Mining (KDD'96) (Palo Alto, CA: AAAI Press), 225
Etzel, P. B. 1981, Photometric and Spectroscopic Binary Systems, ed. E. B. Carling, & Z. Kopal (Dordrecht: Reidel), 111
Gaulme, P., Jackiewicz, J., Appourchaux, T., & Mosser, B. 2014, ApJ, 785, 5
Gaulme, P., McKeever, J., Rawls, M. L., et al. 2013, ApJ, 767, 82
Goldstein, H., Poole, C., & Safko, J. 2002, Classical Mechanics (Reading, MA: Addison-Wesley)
Hadrava, P. 1990, CoSka, 20, 23
Hilditch, R. W. 2001, An Introduction to Close Binary Stars (Cambridge: Cambridge Univ. Press)
Hut, P. 1981, A&A, 99, 126
Kallrath, J., & Milone, E. F. 2009, Eclipsing Binary Stars: Modeling and Analysis (New York: Springer)
Kirk, B., Conroy, K., & Prša, A. 2016, AJ, 151, 68
Kochoska, A., Mowlavi, N., Prša, A., et al. 2017, A&A, 602, A110
Kopal, Z. 1959, Close Binary Systems (New York: Wiley)
Kozai, Y. 1962, AJ, 67, 591
Kwee, K. K., & van Woerden, H. 1956, BAIN, 12, 327
Limber, D. N. 1963, ApJ, 138, 1112
Mardling, R. A. 2013, MNRAS, 435, 2187
Matijevič, G., Prša, A., Orosz, J. A., et al. 2012, AJ, 143, 123
Mazeh, T., & Shaham, J. 1979, A&A, 77, 145
McLaughlin, D. B. 1924, ApJ, 60, 22
Nicholls, C. P., & Wood, P. R. 2012, MNRAS, 421, 2616
Paddock, G. F. 1913, PASP, 25, 208
Prša, A., Batalha, N., Slawson, R.W., et al. 2011, AJ, 141, 83
Prša, A., Guinan, E. F., Devinney, E. J., et al. 2008, ApJ, 687, 542
Richardson, N. D., Moffat, A. F. J., Maltais-Tariant, R., et al. 2016, MNRAS, 455, 244
Roche, É. 1849, La Figure d'une Masse Fluide - Soumise à l'Attraction d'un Point 'Eloigné, part 1, in Académie des Sciences de Montpellier: Mémoires de la section des sciences, Vol. 1, 243
Rossiter, R. A. 1924, ApJ, 60, 15
Roweis, S. T., & Saul, L. K. 2000, Sci., 290, 2323
Rucinski, S. M. 1997, IBVS, 4460, 1
Smart, W. M., & Green, E. b. R. M. 1977, Textbook on Spherical Astronomy (Cambridge: Cambridge Univ. Press)
Sterne, T. E. 1939, MNRAS, 99, 451

Taff, L. G. 1985, Celestial Mechanics: A Computational Guide for the Practitioner (New York: Wiley)
Thompson, S. E., Everett, M., Mullally, F., et al. 2012, ApJ, 753, 86
Thornton, S.T., & Marion, J.B. 2004, Classical Dynamics of Particles and Systems (Pacific Grove, CA: Brooks/Cole)
Van Der Maaten, L., & Hinton, G. E. 2008, JMLR, 9, 2579
Vander Plas, J. T., & Connolly, A. J. 2010, DimReduce: Nonlinear Dimensionality Reduction of Very Large Datasets with Locally Linear Embedding (LLE) and its Variants, Astrophysics Source Code Library, ascl:1010.031
van Hamme, W., & Wilson, R. E. 1986, AJ, 92, 1168
Wilson, R. E. 1979, ApJ, 234, 1054
Wilson, R. E. 2001, IBVS, 5076, 1
Wilson, R. E., & Devinney, E. J. 1971, ApJ, 166, 605
Wilson, R. E., & Sofia, S. 1976, ApJ, 203, 182

Chapter 4

Radiation: The Basics

The computation of fluxes from an extended object has quite a few twists and turns that can trip even a seasoned researcher, so we provide here a step-by-step discussion for the Sun–Earth system in two atmosphere approximations: blackbody and Castelli & Kurucz (2004) model atmosphere. Before we dive into the examples themselves, we need to define several radiative quantities that frame the problem we are trying to solve.

4.1 Intensity

When we measure "light" from a resolved source, we typically measure the amount of energy dE_λ per unit time dt per unit wavelength $d\lambda$ in a beam $d\Omega$ that is radiated from the direction-projected unit area $dA\cos\theta$ (see Figure 4.1):

$$I_\lambda = \frac{dE_\lambda}{dt\; dA\cos\theta\; d\lambda\; d\Omega}. \tag{4.1}$$

This equation defines *intensity*, our first radiative quantity. In physics this quantity is more frequently called spectral radiance. It is worth taking a moment to reflect on the subscript λ in I_λ. This quantity measures the fraction of the overall energy output per unit time that is radiated from the unit area dA in a certain direction, at a certain wavelength interval; thus, it is a *distribution* itself. It tells us how much energy the photons that hit our detector in a certain instant of time will carry with them. This energy can change at the very next instant of time, but we concern ourselves only with the energy carried during the time interval $[t,\ t + dt]$. This leaves the distribution of energy across wavelengths as the primary variable, hence the subscript λ. What we *really* mean, thus, is $I_\lambda \equiv dI/d\lambda$, i.e., a *distribution* of intensities per wavelength interval. If the radiator is a blackbody, this distribution would be prescribed by a Planck function B_λ:

$$B_\lambda = \frac{2hc^2}{\lambda^5}\frac{1}{\exp\left(\dfrac{hc}{\lambda kT}\right) - 1}, \tag{4.2}$$

doi:10.1088/978-0-7503-1287-5ch4

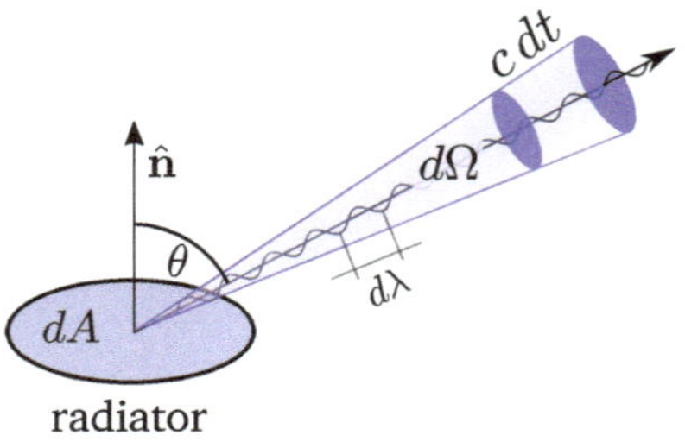

Figure 4.1. Definition of intensity: the amount of energy dE_λ per unit time dt per unit wavelength λ in a beam $d\Omega$ that is radiated from the direction-projected unit area $dA\cos\theta$.

where h is the Planck constant, c is the speed of light, k is the Boltzmann constant, and T is the effective temperature of the radiator. Thus, B_λ is the intensity that corresponds to a blackbody radiator with an effective temperature T. We could also use a sophisticated model that takes into account energy transfer through the outer layers of the star, shuffles photon numbers according to atomic transitions that cause spectral lines, and takes into account opacity variations along photon paths to predict intensity. Such models abound, and one of the well-established, freely accessible codes is that of Kurucz (2013). While the model atmosphere is substantially more complex compared to the Planck function, it still gives us the same quantity of interest: I_λ.

Before we move on, let us consider the units of intensity. It is easiest to do so by inspecting Equation (4.1) directly: we get J s^{-1} m^{-2} m^{-1} (do not let steradians fool you; they are not units), which yields watts per meter *cubed*. This is yet another reminder that we are implicitly referring to a distribution across λ.

4.2 Flux

It is time to introduce our second radiative quantity: flux. By definition, the flux through the radiative element dA is a net sum of all energy per unit time emitted through that element in all directions and projected along the normal of the unit area:

$$F_\lambda = \int_0^{2\pi} I_\lambda \cos\theta d\Omega \equiv \int_0^{\pi/2} \int_0^{2\pi} I_\lambda \cos\theta d\phi \sin\theta d\theta, \tag{4.3}$$

where ϕ and θ are longitudinal and latitudinal components of the solid angle with respect to the normal (i.e., $\theta = 0$ corresponds to the direction along the normal, and ϕ circumscribes the rings around the normal). In physics this quantity is more frequently called spectral flux density. To evaluate this integral, we need to ask ourselves whether I_λ depends on ϕ and/or θ. Strictly speaking, the intensity of real radiators likely does depend on both, but in an idealized case we may be able to get away with less stringent assumptions. In particular, it is reasonable to assume that the amount of energy irradiated along the same angle to the surface element normal is essentially the same, so I_λ will not depend strongly on ϕ. On the other hand, we *do* expect that intensity will depend on the angle to the surface element normal itself, even if we do not know yet how exactly. So for the cylindrically isotropic case, Equation (4.3) can be simplified into:

$$F_\lambda = 2\pi \int_0^{\pi/2} I_\lambda(\theta)\cos\theta \sin\theta d\theta. \tag{4.4}$$

To figure out what drives the dependence of intensity on θ, consider layers upon layers in a stellar atmosphere, stacked like onion shells. Depending on the angle with respect to the surface element normal, the path traversed by a photon trying to escape will be different. The photon's best chance to escape will be along the normal, while any attempts perpendicular to the normal will be much more difficult. Hence, we expect more photons to be emitted along the normal than at an angle. It sounds like this dependence of I_λ on θ should depend on $\cos\theta$ since it is $\cos\theta$ that drives the length of the path taken by the photon through the outermost layer on its path to freedom, but let us be more general than that. First off, let us introduce $\mu = \cos\theta$ to save us from excessive writing, and then let us formulate a function $\mathcal{L}(\mu) = I_\lambda(\mu)/I_{\lambda,\,\mathrm{norm}}$, where $I_{\lambda,\,\mathrm{norm}}$ refers to the intensity along the normal, i.e., $\mu = 1$. If at this point you are tempted to nod and say "ah yes, the limb darkening function," hold that thought: we have not talked about any distribution of unit areas over the surface of the star; we limit our discussion only to the amount of light coming out of a *single* surface element dA. Equation (4.4) can now be rewritten as:

$$F_\lambda = 2\pi \int_1^0 I_\lambda(\mu)\mu(-d\mu) = 2\pi I_{\lambda,\,\mathrm{norm}} \int_0^1 \mu\mathcal{L}(\mu)d\mu. \tag{4.5}$$

Here we used a little algebraic trick $\sin\theta d\theta = -d(\cos\theta) \equiv -d\mu$ and took $I_{\lambda,\,\mathrm{norm}}$ in front of the integral since it does not depend on μ. The negative sign is absorbed by exchanging the integral boundaries. The flux can thus be thought of as normal emergent intensity scaled by $2\pi\int_0^1 \mu\mathcal{L}(\mu)d\mu$. Remember, though, that this has little to do with the flux that we detect as observers elsewhere; this is flux emanating from the unit area on the radiating body.

A closer look at Equation (4.5) can reveal a potential point of confusion: the units of flux appear to be the same as the units of intensity, namely, watts per meter cubed, whereas one might expect the units to be watts per meter *squared*. Do not let the established sloppiness in referring to flux density as "flux" confuse you: by F_λ we mean the *distribution* of flux across wavelengths, which is sometimes referred to as monochromatic flux. The actual flux is obtained by integrating over wavelength, and that will sort out the apparent units conundrum.

Now let us step away from the radiating body and place an observer at some arbitrary point at a distance D from the radiating unit area. The amount of energy radiated toward the observer is given by Equation (4.1):

$$dE_\lambda = I_\lambda(\theta)\, dt\, dA \cos\theta\, d\lambda\, d\Omega. \tag{4.6}$$

By the same equation, the amount of energy intercepted at the observer's point by a projected unit area dA' is:

$$dE'_\lambda = I'_\lambda(\theta')\, dt\, dA' \cos\theta'\, d\lambda\, d\Omega'. \tag{4.7}$$

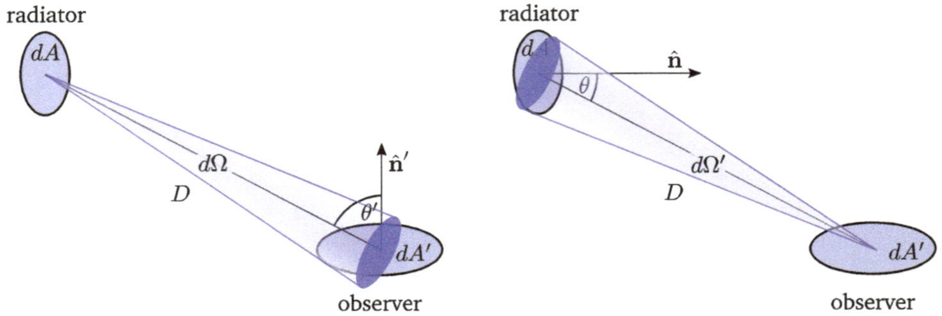

Figure 4.2. Different orientations of $d\Omega$ and $d\Omega'$ explain why $d\Omega$ is proportional to $A'\cos\theta'$ and $d\Omega'$ is proportional to $A\cos\theta$, i.e., Equation (4.8).

Note that $d\Omega$ and $d\Omega'$ are directed in the opposite sense: $d\Omega$ is a cone from the radiator to the observer, while $d\Omega'$ is a cone from the observer to the radiator (see Figure 4.2). If there is no attenuating matter between the emitting and intercepting unit areas, there is nothing to take away the energy, so it must be conserved: $dE_\lambda = dE'_\lambda$. At the same time, the solid angle differential $d\Omega$ depends on the distance and the projected intercepting area, while the solid angle differential $d\Omega'$ depends on the distance and the projected emitting area (as depicted in Figure 4.2):

$$d\Omega = \frac{dA'_\perp}{D^2} \equiv \frac{dA'\cos\theta'}{D^2}, \qquad d\Omega' = \frac{dA_\perp}{D^2} \equiv \frac{dA\cos\theta}{D^2}. \tag{4.8}$$

Plugging these two expressions into Equations (4.6)–(4.7) immediately leads to two important conclusions: first, $I_\lambda(\theta) = I'_\lambda(\theta')$, i.e., the intensity of the beam is conserved and it does not depend on the distance, and second, $\mathcal{L}(\theta) = \mathcal{L}'(\theta')$, i.e., *now* it is appropriate to recognize that $\mathcal{L}(\theta)$ is indeed the limb-darkening function. This realization also allows us to compute the flux originating at the emitting unit area dA and flowing into the intercepting unit area dA' at the observer's location (see Figure 4.3):

$$dF_\lambda = I'_\lambda(\theta')\cos\theta' d\Omega' \equiv I_\lambda(\theta)\frac{\cos\theta'\cos\theta}{D^2}dA. \tag{4.9}$$

If the intercepting unit area is perpendicular to the beam (i.e., a detector pointing toward the light source), $\cos\theta' = 1$.

Suppose we now put a filter on the detector at the observer's location. This filter transmits only a part of the overall radiation and blocks the rest. Let its transmittance be described by a passband function $\mathcal{P}(\lambda)$. This reduces the amount of monochromatic flux on the detector to:

$$dF_\lambda = \mathcal{P}(\lambda)I_\lambda(\theta)\frac{\cos\theta'\cos\theta}{D^2}dA, \tag{4.10}$$

which we can interpret as $\mathcal{P}(\lambda)$ being a "sifting" function that determines how much of the intercepted flux on the $[\lambda, \lambda + d\lambda]$ interval actually reaches the detector. Any

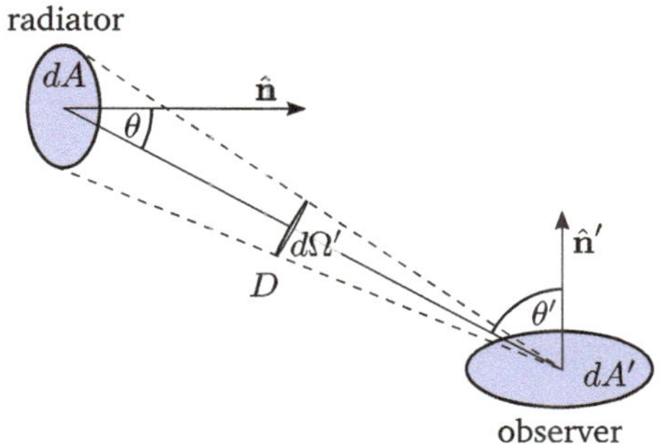

Figure 4.3. Amount of flux dF_λ intercepted at the observer's location.

real optics/detector system will cause further loss of radiation because of the inevitable efficiency limitations, and we can formally bundle all such effects into $\mathcal{P}(\lambda)$.

There is another consideration worth mentioning at this point: not all detectors are created equal in terms of what they measure. We are usually slack and say that the detector measures "light," but by that we mean either *energy*, which we have discussed so far, or *photon counts*. Depending on the quantity measured, detectors are either bolometric (measuring energy) or photoelectric (counting photons). Bolometric detectors have no notion of the number of photons detected, only their net energy; conversely, photoelectric detectors have no notion of energy, only the net number of photons. An example of a bolometric detector is a photoelectric photomultiplier. An example of a photoelectric detector is a charge coupled device (CCD). Because of this distinction, our equation for flux needs to be suitably adapted for the photon-counting instruments. The energy carried by a single photon on the wavelength interval $[\lambda, \lambda + d\lambda]$ is given by hc/λ, so the net energy at that wavelength interval is $dE_\lambda = dN_\lambda\, hc/\lambda$, where dN_λ is the number[1] of photons with wavelengths between λ and $\lambda + d\lambda$. Thus, if intensity is defined in terms of dN_λ instead of dE_λ, as appropriate for the photon-counting detectors, the expression for the passband flux becomes:

$$dF_\lambda^{\mathrm{pc}} = \frac{1}{hc}\,\lambda\mathcal{P}(\lambda)I_\lambda(\theta)\,\frac{\cos\theta'\cos\theta}{D^2}dA. \tag{4.11}$$

Here we used superscript "pc" to denote a photon-counting flux. Our realization from before that $\mathcal{L}(\theta) \equiv \mathcal{L}'(\theta')$ also allows us to express the monochromatic flux from the radiating unit area in the following form:

$$\begin{aligned} dF_\lambda &= \mathcal{P}(\lambda)I_{\lambda,\,\mathrm{norm}}\mathcal{L}(\mu)\frac{\mu\mu'}{D^2}dA, \\ dF_\lambda^{\mathrm{pc}} &= \frac{1}{hc}\,\lambda\mathcal{P}(\lambda)I_{\lambda,\,\mathrm{norm}}\mathcal{L}(\mu)\frac{\mu\mu'}{D^2}dA, \end{aligned} \tag{4.12}$$

[1] In our established practice of "slack" terminology, dN_λ is actually the *distribution* of the number photons per wavelength interval, indicated by the λ subscript.

where, as stated before, $\mu' \equiv \cos\theta' = 1$ in most[2] practical purposes.

Equations (4.12) set the stage for integration over the visible area of an extended body and over all wavelengths into the actual observed flux in absolute units (watts per meter squared). In the discussion that follows, we will derive expressions for bolometric detectors, but the results will be easily generalized to photon-counting devices as well.

In general, an extended body can have a complex shape, and parts of it may eclipse other parts; to describe this, let us introduce a visibility function $\mathcal{V}(\boldsymbol{r})$, where $\boldsymbol{r}$ gives the location on the radiating body surface, as:

$$\mathcal{V}(\boldsymbol{r}) = \begin{cases} 1 & \text{if } \mathrm{dA}(\boldsymbol{r}) \text{ is visible,} \\ 0 & \text{otherwise.} \end{cases} \tag{4.13}$$

The integral over the visible area can then still be performed over the entire surface ∂V of the radiating body with volume V,

$$F_\lambda = \frac{\mu'}{D^2}\mathcal{P}(\lambda)\int_{\partial V} \mu\mathcal{V}(\boldsymbol{r})I_{\lambda,\,\mathrm{norm}}\mathcal{L}(\mu)dA, \tag{4.14}$$

and the integral over all wavelengths (remember, $F_\lambda \equiv dF/d\lambda$) then yields the total flux:

$$F = \frac{\mu'}{D^2}\int_0^\infty \mathcal{P}(\lambda)\int_{\partial V} \mu\mathcal{V}(\boldsymbol{r})I_{\lambda,\,\mathrm{norm}}\mathcal{L}(\mu)dAd\lambda. \tag{4.15}$$

4.3 Luminosity

The final radiative quantity of interest to the current discussion is *luminosity*. By definition, luminosity is given by the net flux over all wavelengths over the radiating surface area:

$$L = \int_{\partial V}\int_{\lambda=0}^\infty F_\lambda d\lambda dA. \tag{4.16}$$

The units of luminosity are watts, implying that, indeed, luminosity is equivalent to what we would call *power* in physics. Note that luminosity is a property of the radiating body that is entirely independent of the position (or even the existence) of the observer: it measures the net energy output, and it will not change if it is near or far, or if something blocks it or reflects its light. It is an intrinsic property of the radiating body. Yet it still makes sense to ask what would be the luminosity of a radiating body in a given passband. It is crucially important not to think of this as a luminosity that somehow depends on the observational setup; instead, the question is how much power is in the spectral range defined by the $\mathcal{P}(\lambda)$ function. We call this passband luminosity and express it as:

$$L_{\mathrm{pb}} = \int_{\partial V}\int_0^\infty 2\pi\mathcal{P}(\lambda)I_{\lambda,\,\mathrm{norm}}\left(\int_0^1 \mu\mathcal{L}(\mu)d\mu\right)d\lambda dA. \tag{4.17}$$

[2] One notable exception is the treatment of reflection.

There are two important points to make here. First, bolometric luminosity can be considered a passband luminosity where the "passband" is simply $\mathcal{P}(\lambda) = 1$ for all λ. Thus, we can use $L_{\rm pb}$ for any luminosity, including bolometric, without any loss of generality. Second, it might seem surprising that luminosity depends on the limb-darkening function $\mathcal{L}(\mu)$; in this case, $\mathcal{L}(\mu)$ refers not to limb darkening but to the variation of intensity from the radiating unit area with direction μ; the equivalence with limb darkening stems from the conservation of intensity stemming from Equation (4.8). As we need to consider *all* radiated light, we certainly expect the dependence on $\mathcal{L}(\mu)$ or, more precisely, its integral over all μ.

4.4 Limb Darkening

So far we have made a few ad hoc references to the limb-darkening function $\mathcal{L}(\mu)$ without formally introducing the processes that shape its dependence on μ. As mentioned before, emergent intensity $I_\lambda(\mu)$ depends on the conditions in the stellar atmosphere. As light propagates outward, a fraction of it is absorbed by way of bound–bound transitions, bound–free absorption (photoionization), free–free absorption, and electron scattering. This extinction can be described by:

$$dI_\lambda = -\kappa_\lambda \rho I_\lambda dl, \tag{4.18}$$

where κ_λ is the absorption coefficient (or opacity), ρ is the local gas density, and dl is the path differential. The absorption coefficient κ_λ is generally a function of the wavelength, composition, density, and temperature of the gas: $\kappa_\lambda = \kappa_\lambda([\mathrm{M/H}], \rho, T_{\rm eff})$.

At the same time, some of the light is added to intensity by way of scattering, bound–bound emission, and recombination:

$$dI_\lambda = j_\lambda \rho dl, \tag{4.19}$$

where j_λ is the emission coefficient of the gas that generally depends on the same conditions as κ_λ. Note one important distinction, though: the contribution of emission to the emergent intensity is independent of the intensity I_λ itself. The reason is that this extra emission comes from the surrounding gas and thus has nothing to do with intensity itself.

The *net* change in emergent intensity is the sum of both emission and absorption:

$$dI_\lambda = j_\lambda \rho dl - \kappa_\lambda \rho I_\lambda dl. \tag{4.20}$$

The ratio of the rates at which the competing processes of emission and absorption occur determines how rapidly the intensity changes as it propagates layer after layer through the atmosphere. Dividing Equation (4.20) by $(-\kappa_\lambda \rho dl)$ yields:

$$-\frac{1}{\kappa_\lambda \rho}\frac{dI_\lambda}{dl} = I_\lambda - \frac{j_\lambda}{\kappa_\lambda} \equiv I_\lambda - \mathcal{S}_\lambda, \tag{4.21}$$

where $\mathcal{S}_\lambda = j_\lambda/\kappa_\lambda$ is called the source function and the equation in this form is called the *equation of radiative transfer*. The term $1/\kappa_\lambda \rho$ has a special significance: it determines a typical attenuation length[3] and is called the *mean free path*. We can now define *optical depth* as $d\tau_\lambda = -\kappa_\lambda \rho dl$, where the negative sign implies that we measure it from the top of the stellar atmosphere inward. Based on the optical depth of the material, we can define its optical *thickness*: for $\tau_\lambda \gg 1$ the gas is said to be optically *thick*, while for $\tau_\lambda \ll 1$ it is said to be optically *thin*. Using optical depth, Equation (4.21) can be written as:

$$\frac{dI_\lambda}{d\tau_\lambda} = I_\lambda - \mathcal{S}_\lambda. \tag{4.22}$$

This is a nonhomogeneous differential equation of the first order. If the dependence of the source function $\mathcal{S}_\lambda$ on the optical depth τ_λ were known, Equation (4.22) could be solved to find emergent intensity as a function of μ. This is what model atmospheres do: they predict, using complex machinery, $\mathcal{S}_\lambda$ throughout the stellar photosphere. Simplified approximations for $\mathcal{S}_\lambda$ exist that allow analytical treatment, but at this point we will forgo the mathematical rigor in deriving the explicit relationship between the emergent intensity and an assumed modeling function for $\mathcal{S}_\lambda$ because it is lengthy and not really all that important for the discussion; instead, let us qualitatively review the main points of the process and refer those interested in full derivation to Gray (2008).

First, it is not immediately evident where μ enters Equation (4.22). Recall that $d\tau_\lambda = -\kappa_\lambda \rho dl$, i.e., the optical depth refers to opacity along the path *dl*, *whatever* the direction of *dl* may be. If we were to write dI_λ/dl in polar coordinates, for example, it would be:

$$\frac{dI_\lambda}{dl} = \frac{\partial I_\lambda}{\partial r}\frac{dr}{dl} + \frac{\partial I_\lambda}{\partial \theta}\frac{d\theta}{dl}. \tag{4.23}$$

In cases where the thickness of the visible atmosphere is small compared to the size of the radiator, we can neglect the second term on the right (i.e., we assume the plane-parallel approximation). At the same time, dr/dl is a projection of the path onto the radius vector, which is precisely μ—i.e., the dependence we were looking for—rendering the transfer equation to:

$$\mu\frac{dI_\lambda}{d\tau_{\lambda,\mu}} = I_\lambda - \mathcal{S}_\lambda, \tag{4.24}$$

where we denote with $\tau_{\lambda,\mu}$ the optical depth along the line of sight determined by μ.

Next, we need to ask ourselves how "deep" into the star we can see. The optical depth at the top of the atmosphere is 0, and it increases gradually toward the center of the star. The *photosphere* is a part of the overall stellar atmosphere from where the light we see originates; hence, it determines the effective temperature and the size of the star. This depth generally delimits the outer region of the star where

[3] This can be seen easily from the solution of Equation (4.18), $I_\lambda = I_{\lambda,0}\exp(-\kappa_\lambda \rho L)$.

photons scatter on average less than once on their way out. Even though formally we need to integrate τ_λ all the way to infinity, the exponentially growing optical depth ascertains rapid convergence of the integral in Equation (4.24). This is why the Sun does not appear "fuzzy"—the photosphere is quite thin compared to the star itself. Eddington has shown that, under the assumption of a gray atmosphere (i.e., κ_λ and j_λ do not depend on λ), the photosphere base is at an optical depth of 2/3. The actual depth will depend on the wavelength, of course—that is why we can "x-ray" the Sun. We generally consider that, in the optical, the photosphere is essentially opaque at $\tau_\lambda \sim 1$.

To compute emergent intensity in a given direction, we then need to integrate Equation (4.24) from the top of the photosphere to the depth at which opacity makes the interior opaque. This will be the same *optical* depth but a different *physical* depth depending on the angle to the surface normal we look at (see Figure 4.4). If we are looking at the center, the optical depth path is antiparallel to the radius and we will see the deepest, hottest, brightest layers. If we are looking toward the limb, the optical depth path is nearly perpendicular and we will see the shallowest, coolest, dimmest layers. In effect, we will see surface brightness variation because $\mathcal{S}_\lambda$ is different for each photosphere layer. It is this surface brightness variation that we see as limb darkening. We parameterize it by describing $I_\lambda(\mu)/I_{\lambda,\,\mathrm{norm}}$ as a function of μ. There are several models commonly used today; they are summarized in Table 4.1.

Note that the plane-parallel approximation embedded in Castelli & Kurucz (2004) model atmospheres breaks down for giant and supergiant stars. The effects of surface curvature are pronounced, and advanced treatment of atmospheres is necessary. There are several models that take curvature into account, for example, Phoenix/NextGen (Hauschildt et al. 1999) and SAtlas (Lester & Neilson 2008). NextGen models have been successfully integrated into the binary star modeling code ELC (Orosz & Hauschildt 2000). The curvature significantly affects the limb darkening, and the coefficients interpolated from the Castelli & Kurucz (2004) tables will likely be inadequate for giants and supergiants. Tables computed from spherical models, i.e., by Claret & Bloemen (2011), Neilson & Lester (2013), and similar works, should be considered instead.

4.5 Computational Efficiency

We now digress for a moment to address an important computational aspect of Equations (4.15) and (4.17). Recall that $\mathcal{P}(\lambda)I_{\lambda,\,\mathrm{norm}}(\lambda)$ corresponds to the passband-attenuated monochromatic intensity in the direction along the normal of the radiating unit area; if we integrate this expression over all wavelengths, we will get a normal emergent passband intensity:

$$I_{\mathrm{pb,\,norm}} = \int_0^\infty \mathcal{P}(\lambda)I_{\lambda,\,\mathrm{norm}}(\lambda)d\lambda \equiv \int_{\lambda_{\mathrm{lower}}}^{\lambda_{\mathrm{upper}}} \mathcal{P}(\lambda)I_{\lambda,\,\mathrm{norm}}(\lambda)d\lambda, \tag{4.25}$$

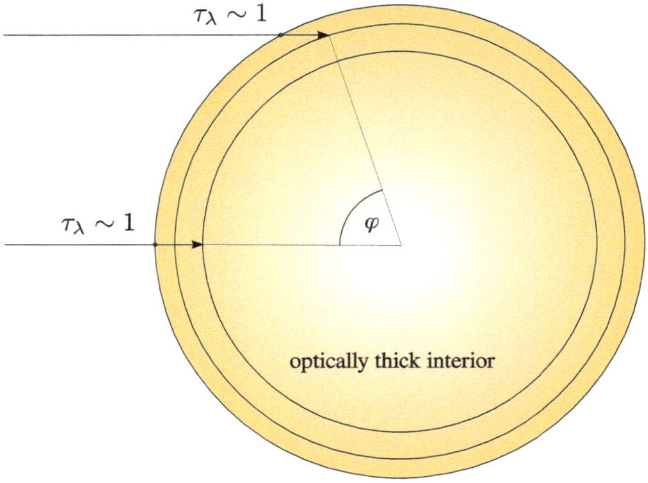

Figure 4.4. Limb-darkening effect. A common way to analytically quantify the optical depth of the visible photospheric surface is to adopt a model of a plane-parallel atmosphere. When the atmosphere is assumed to be gray (i.e., κ_λ is independent of λ), the reference surface by Eddington's approximation corresponds to the layer at optical depth $\tau_\lambda = 2/3$. This is a limited approximation because all sources of opacity except for Thompson scattering are wavelength dependent, but in general the reference layer will be at the optical depth of ~1.

Table 4.1. Table of Commonly Used Limb-darkening Models

Model	Functional Form	References
Linear	$\mathcal{L}(\mu) = 1 - x_\lambda(1 - \mu)$	1
Logarithmic	$\mathcal{L}(\mu) = 1 - x_\lambda(1 - \mu) - y_\lambda \mu \log(\mu)$	2
Square root	$\mathcal{L}(\mu) = 1 - x_\lambda(1 - \mu) - y_\lambda(1 - \sqrt{\mu})$	3
Quadratic	$\mathcal{L}(\mu) = 1 - x_\lambda(1 - \mu) - y_\lambda(1 - \mu)^2$	4
Exponential	$\mathcal{L}(\mu) = 1 - x_\lambda(1 - \mu) - y_\lambda(1 - \exp \mu)^{-1}$	5
Power	$\mathcal{L}(\mu) = 1 - a_{\lambda,1}(1 - \mu^{1/2}) - a_{\lambda,2}(1 - \mu) - a_{\lambda,3}(1 - \mu^{3/2}) - a_{\lambda,4}(1 - \mu^2)$	6
Interpolated	$\mathcal{L}(\mu) = \text{interp}\,(\mu;\, T_{\text{eff}},\, \log g,\, [\text{M/H}])$	7

References. (1) Schwarzschild 1906; (2) Klinglesmith & Sobieski 1970; (3) Diaz-Cordoves & Gimenez 1992; (4) Kopal 1950; (5) Claret & Hauschildt 2003; (6) Claret 2000; (7) Prša et al. 2016b.

where λ_{lower} and λ_{upper} are the limiting wavelengths of $\mathcal{P}(\lambda)$, i.e., the wavelengths below which and above which $\mathcal{P}(\lambda) = 0$, respectively. It proves computationally more practical to work with the *mean* normal passband intensity, computed as:

$$\langle I_{\text{pb, norm}} \rangle = \frac{\int_\lambda \mathcal{P}(\lambda) I_{\lambda,\,\text{norm}}(\lambda) d\lambda}{\int_\lambda P(\lambda) d\lambda} \equiv \frac{I_{\text{pb, norm}}}{\int_\lambda P(\lambda) d\lambda}. \tag{4.26}$$

Plugging this into Equation (4.15) yields the passband flux,

$$F_{\mathrm{pb}} = \frac{\mu'}{D^2}\left(\int_\lambda \mathcal{P}(\lambda)d\lambda\right)\int_{\partial V}\langle I_{\mathrm{pb,\,norm}}\rangle\mu\mathcal{V}(\boldsymbol{r})\mathcal{L}(\mu)dA, \tag{4.27}$$

and an equivalent expression for the photon-counting devices,

$$F_{\mathrm{pb}}^{\mathrm{pc}} = \frac{\mu'}{D^2}\left(\frac{1}{hc}\int_\lambda \lambda\mathcal{P}(\lambda)d\lambda\right)\int_{\partial V}\langle I_{\mathrm{pb,\,norm}}^{\mathrm{pc}}\rangle\mu\mathcal{V}(\boldsymbol{r})\mathcal{L}(\mu)dA. \tag{4.28}$$

The reason why Equations (4.27)–(4.28) are more practical than Equation (4.15) is that the mean normal passband intensities $\langle I_{\mathrm{pb,\,norm}}\rangle$ and $\langle I_{\mathrm{pb,\,norm}}^{\mathrm{pc}}\rangle$ can be computed ahead of time from model atmospheres, and $\int_\lambda \mathcal{P}(\lambda)d\lambda$ and $(hc)^{-1}\int_\lambda \lambda\mathcal{P}(\lambda)d\lambda$ depend only on the transmission functions, so they can be computed in advance as well.

We can do the same for the luminosity computation:

$$L_{\mathrm{pb}} = \left(\int_0^\infty \mathcal{P}(\lambda)d\lambda\right)\int_{\partial V} 2\pi\langle I_{\mathrm{pb,\,norm}}\rangle\left(\int_0^1 \mu\mathcal{L}(\mu)d\mu\right)dA, \tag{4.29}$$

or, for the photon-counting detectors,

$$L_{\mathrm{pb}}^{\mathrm{pc}} = \left(\frac{1}{hc}\int_0^\infty \lambda\mathcal{P}(\lambda)d\lambda\right)\int_{\partial V} 2\pi\langle I_{\mathrm{pb,\,norm}}^{\mathrm{pc}}\rangle\left(\int_0^1 \mu\mathcal{L}(\mu)d\mu\right)dA. \tag{4.30}$$

In conclusion, the radiation in absolute units can thus be described by two pairs of equations, summarized in Table 4.2.

Turning to limb darkening, it is impractical to include the computation of limb-darkening coefficients in light-curve generators on the fly because of the complexity and numerical time cost of stellar atmosphere models. We instead resort to precomputed lookup tables. These are generated by synthesizing $I_\lambda(\mu)$ in absolute units from model atmospheres across the valid ranges of T_{eff}, $\log g$, [M/H], and possibly other relevant atmospheric parameters and computing $I_\lambda(\mu)/I_{\lambda,\,\mathrm{norm}}$ for each set of parameters. We then fit any of the laws in Table 4.1 to these data to obtain the best-fit coefficients. Figure 4.5 depicts an example of a Sun-like limb darkening (filled circles), with fits for several limb-darkening models overplotted. It is clear that one- and even two-parameter models are not adequate for high-precision modeling.

Table 4.2. Equations That Govern the Computation of Flux from and Luminosity of a Radiating Body

Parameter	Equations
Passband flux	$F_{\mathrm{pb}} = \frac{\mu'}{D^2}(\int_\lambda \mathcal{P}(\lambda)d\lambda)\int_{\partial V}\langle I_{\mathrm{pb,\,norm}}\rangle\mu\mathcal{V}(\mathbf{r})\mathcal{L}(\mu)dA$
	$F_{\mathrm{pb}}^{\mathrm{pc}} = \frac{\mu'}{D^2}(\frac{1}{hc}\int_\lambda \lambda\mathcal{P}(\lambda)d\lambda)\int_{\partial V}\langle I_{\mathrm{pb,\,norm}}^{\mathrm{pc}}\rangle\mu\mathcal{V}(\mathbf{r})\mathcal{L}(\mu)dA$
Passband luminosity	$L_{\mathrm{pb}} = (\int_0^\infty \mathcal{P}(\lambda)d\lambda)\int_{\partial V} 2\pi\langle I_{\mathrm{pb,\,norm}}\rangle(\int_0^1 \mu\mathcal{L}(\mu)d\mu)dA$
	$L_{\mathrm{pb}}^{\mathrm{pc}} = (\frac{1}{hc}\int_0^\infty \lambda\mathcal{P}(\lambda)d\lambda)\int_{\partial V} 2\pi\langle I_{\mathrm{pb,\,norm}}^{\mathrm{pc}}\rangle(\int_0^1 \mu\mathcal{L}(\mu)d\mu)dA$

Note. Terms in angle brackets and in parentheses can be computed ahead of time to make the computation efficient.

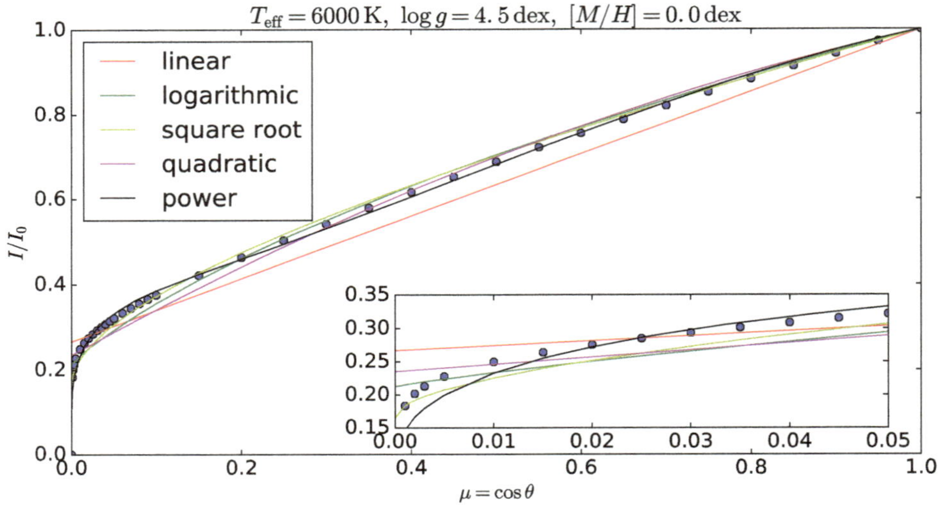

Figure 4.5. Variation of emergent passband intensity (with respect to normal emergent passband intensity) with emergent angle $\mu \equiv \cos\theta$. The inset shows a zoomed-in region at low μ. Filled circles are integrated intensities based on Castelli & Kurucz (2004) model atmospheres, and lines correspond to the fitted limb-darkening models using least squares. Systematics that arises from using these low-parametric models can be significant.

Further cuts in the parameter space spun by limb-darkening coefficients based on physical considerations can be made efficiently (Kipping 2013; Espinoza & Jordán 2016), but ultimately the accuracy of limb-darkening treatment will be limited either by the quality of the model fit or by the systematics in model atmospheres.

4.6 Relative Units

Our discussion so far has pertained to intensities, fluxes, and luminosities expressed in absolute physical units. However, in many use cases in astronomy this is not warranted or even possible, i.e., when we only have differential photometry, or we do not know the distance to the object, or we do not know the object's effective temperature accurately, or any combination of these and other limitations. The ongoing *Gaia* mission (Brown et al. 2016) holds the promise to change that by providing us with microarcsecond parallaxes, but until then, a common case calls for *relative* instead of absolute units.

What do the relative units entail? By relative we mean to say that whatever units are implied by the passband luminosity, those units are carried over to other radiative quantities. These can still be physical, of course, but more often they are referential, i.e., they yield a flux that allows direct comparison with observations that can be scaled[4] more or less arbitrarily. By *assigning* a certain passband luminosity,

[4] A common practice is to flux-normalize the data at some reference time stamp and use that as a point of reference that determines passband luminosity.

we are essentially imposing a scaling $\xi = L_{\rm pb}^{\rm rel}/L_{\rm pb}^{\rm abs}$ that satisfies the following relationship:

$$\begin{aligned} L_{\rm pb}^{\rm rel} &= \left(\int_0^\infty \mathcal{P}(\lambda)d\lambda\right)\int_{\partial V} 2\pi\langle I_{\rm pb,\,norm}^{\rm rel}\rangle\left(\int_0^1 \mu\mathcal{L}(\mu)d\mu\right)dA \\ \xi L_{\rm pb}^{\rm abs} &= \xi\left(\int_0^\infty \mathcal{P}(\lambda)d\lambda\right)\int_{\partial V} 2\pi\langle I_{\rm pb,\,norm}^{\rm abs}\rangle\left(\int_0^1 \mu\mathcal{L}(\mu)d\mu\right)dA, \end{aligned} \tag{4.31}$$

from which it follows directly that $\xi = \langle I_{\rm pb,\,norm}^{\rm rel}\rangle/\langle I_{\rm pb,\,norm}^{\rm abs}\rangle$. Thus, all equations in Table 4.2 retain their form; they are only multiplied by ξ to transform to the relative units suitably.

When we want to use photon counts to correctly weigh the data gathered by a photon-counting device but want to express the values in energy units, it helps to realize that:

$$L_{\rm pb} = F_{\rm pb}A, \quad L_{\rm pb}^{\rm pc} = F_{\rm pb}^{\rm pc}A \quad \Rightarrow \quad \frac{L_{\rm pb}^{\rm pc}}{L_{\rm pb}} = \frac{F_{\rm pb}^{\rm pc}}{F_{\rm pb}}. \tag{4.32}$$

This holds true for both absolute and relative units because the scaling factor ξ is canceled out. Thus, if ξ is the correct scaling factor for energy-weighted intensities, the same factor will scale the photon-weighted intensities as well.

4.7 Blackbody Radiation

The fundamental assumption of blackbody radiation is local thermodynamic equilibrium (LTE) of the radiator itself: whatever net energy flows into any part of the radiator, the same net energy leaves that part of the radiator, implying that the radiation within any unit volume of a blackbody radiator is *isotropic*. It is not just cylindrically isotropic, as was the case in Equation (4.4); it is *truly* isotropic, in all directions. For a blackbody radiator that means that whatever energy flows from the interior to the surface will be emitted to the exterior, none will be reflected back, and the amount of energy emitted will depend only on the effective temperature. Just *how* that energy will be radiated out is governed by the fact that the radiation inside is isotropic, implying that the radiation outward must also be isotropic. This has a fundamental implication: from whichever direction we look at a unit area of a blackbody radiator, we will see the same intensity. In other words, $I_\lambda(\mu)$ is equal to $I_{\lambda,\,\rm norm}$ for all μ, implying $\mathcal{L}(\mu) \equiv 1$:

$$F_\lambda = 2\pi I_{\lambda,\,\rm norm}\int_0^1 \mu d\mu = \pi I_{\lambda,\,\rm norm} \equiv \pi B_\lambda, \tag{4.33}$$

where $B(\lambda)$ is the Planck function given by Equation (4.2). This is known as Lambert's cosine law, where the "cosine" part refers to the term $\mu \equiv \cos\theta$ inside the integral. Integrating Equation (4.33) over all wavelengths then yields total flux:

$$F = \pi\int_0^\infty B_\lambda d\lambda = \pi\frac{\sigma}{\pi}T^4 \equiv \sigma T^4. \tag{4.34}$$

Here $\sigma = 2\pi^5 k^4/15h^3c^2$ is the Stefan–Boltzmann constant, and Equation (4.34) is known as Stefan's law.

The source function (Equation (4.22)) of a blackbody radiator is also of notable interest: it is equal to the Planck function B_λ, given by Equation (4.2). This is because pure emission and pure absorption of a blackbody are exactly balanced. As I_λ is constant, $dI_\lambda/dl = 0$ and $\mathcal{S}_\lambda = I_\lambda = B_\lambda$. This is important for stellar atmospheres, since the blackbody assumption of LTE immediately implies that the source function $S_\lambda = B_\lambda$.

To get blackbody flux in terms of photon counts instead, we do the same as before: we replace dE_λ with dN_λ in the expression for intensity, from which it immediately follows that $I_\lambda^{\mathrm{pc}} = \lambda I_\lambda/hc$ and, from Equation (4.34),

$$F^{\mathrm{pc}} = \frac{\pi}{hc}\int_0^\infty \lambda B_\lambda d\lambda = 4\pi c\left(\frac{k}{hc}\right)^3 \zeta(3) T^3, \tag{4.35}$$

where $\zeta(3) = 1.2020569\cdots$ is the Riemann zeta function. Combining Equations (4.34) and (4.35) allows us to express F^{pc} with F as:

$$\begin{aligned} F^{\mathrm{pc}} &= 4\pi c\left(\frac{k}{hc}\right)^3 \frac{\zeta(3)}{\sigma^{3/4}} F^{3/4} \\ &= (4.137774 \times 10^{20}\mathrm{s}^{-1}\mathrm{m}^{-1/2}\mathrm{W}^{-3/4}) F^{3/4}. \end{aligned} \tag{4.36}$$

Analogously, the relationship between the luminosity in photon counts per second and luminosity in watts can be derived by observing that $L^{\mathrm{pc}}/L = F^{\mathrm{pc}}/F$, from which it immediately follows that:

$$L^{\mathrm{pc}} = 4\pi c\left(\frac{k}{hc}\right)^3 \frac{\zeta(3)}{\sigma^{3/4}} A^{1/4} L^{3/4}, \tag{4.37}$$

where A is the surface area of the radiating body.

There is another subtlety with blackbody radiators that we need to take into account: we need to ensure consistency across luminosity and flux in the presence of limb darkening. This is tricky because, according to Lambert's law (Equation (4.33)), a blackbody disk would appear uniformly bright, i.e., there would be no inherent limb darkening. Stars, of course, are limb darkened, so any $\mathcal{L}(\mu) \neq 1$ applied to blackbody fluxes will cause the luminosity to change. This is a problem because $I_{\lambda,\,\mathrm{norm}}$ depends solely on the temperature (Equation (4.2)), and when integrated over the surface area of a star, it yields luminosity (Equation (4.16)) that is independent of any limb darkening:

$$L = \int_{\partial V}\int_0^\infty F_\lambda d\lambda dA = \int_{\partial V} \sigma T^4 dA = \sigma T^4 A. \tag{4.38}$$

When we apply the limb-darkening correction, we reduce the flux that corresponds to the effective temperature, thus causing a discrepancy between the actual effective temperature and the "apparent" effective temperature. Say, for example, that $\mathcal{L}(\mu) = 0.5$ for all μ: that would reduce F_λ by a factor of 2 across the board, implying that the "apparent" effective temperature is a factor of $2^{1/4}$ smaller than the actual effective temperature—hence the discrepancy. To ensure consistency between luminosity and flux, it is $I_{\lambda,\,\mathrm{norm}}$ in Equation (4.5) that needs to increase by the amount that limb darkening decreases the flux. In other words, Equation (4.5) for blackbody radiators becomes:

$$F_{\lambda,\,\mathrm{bb}} = 2\pi \left(\frac{I_{\lambda,\,\mathrm{norm}}}{\int_0^1 \mu \mathcal{L}(\mu) d\mu} \right) \int_0^1 \mu \mathcal{L}(\mu) d\mu, \tag{4.39}$$

where the quantity in parentheses is our effective monochromatic blackbody intensity, which is larger than the Planckian intensity by the overall amount reduced by limb darkening. When integrated over the surface, this will yield the luminosity that corresponds to the actual effective temperature, not to the "apparent" effective temperature.

4.8 Putting It All Together

Equipped with the framework developed so far, we can now dive into our example: the Sun–Earth system within the blackbody approximation and with model atmospheres. The reader is strongly encouraged to take this opportunity and work through the examples with a calculator in hand.

We start by assigning the following numerical values to the Sun (Prša et al. 2016a):

Luminosity:	$L_\odot = 1\,\mathcal{L}_\odot^{\mathrm{N}} \equiv 3.823 \times 10^{26}$ W
Radius:	$R_\odot = 1\,\mathcal{R}_\odot^{\mathrm{N}} \equiv 6.957 \times 10^{8}$ m
Effective temperature:	$T_\odot = 1\,\mathcal{T}_\odot^{\mathrm{N}} \equiv 5.772 \times 10^{3}$ K
Synchronicity parameter:	$F_\odot = 14.61$

and the following numerical values to the orbit:

Mass ratio:	$q = \mathcal{GM}_{\mathrm{E}}^{\mathrm{N}} / \mathcal{GM}_\odot^{\mathrm{N}} = 3.003489 \times 10^{-6}$
Semimajor axis:	$a = 149,\,597,\,870,\,700$m

Let us first compute analytical reference values for a spherical Sun that we will use as gauges for the numerical computation. The bolometric flux at Earth is:

$$F_{\mathrm{bol}} = \frac{\mathcal{L}_\odot^{\mathrm{N}}}{4\pi a^2} = \frac{3.828 \times 10^{26}\ \mathrm{W}}{4\pi (149,\,597,\,870,\,700\ \mathrm{m})^2} = 1361.166\ \mathrm{Wm}^{-2}. \tag{4.40}$$

Using Equations (4.36) and (4.37), we can readily obtain bolometric radiative quantities in photon counts:

$$\begin{aligned} F_{\rm bol}^{\rm pc} &= 6.323 \times 10^{21}\ \text{photons m}^{-2}\,\text{s}^{-1}, \\ L_{\rm bol}^{\rm pc} &= 1.778 \times 10^{45}\ \text{photons s}^{-1}. \end{aligned} \tag{4.41}$$

Now let us investigate how close we can get to the bolometric values by using a quasi-bolometric box passband:

$$\mathcal{P}(\lambda) = \begin{cases} 1 & \text{for } 90\ \text{nm} \leqslant \lambda \leqslant 4\ \mu\text{m} \\ 0 & \text{otherwise.} \end{cases} \tag{4.42}$$

This wavelength region is chosen because the list of spectral lines is most complete in this region, which will allow us to make a one-to-one comparison with model atmospheres later on. Thanks to our spherical Sun approximation, the equations in Table 4.2 are significantly simplified:

$$\begin{aligned} F_{\rm pb} &= \Delta\lambda \langle I_{\rm pb,\,norm}\rangle \pi \mathcal{R}_\odot^{\rm N^2}/D^2 = 1347.879\ \text{Wm}^{-2}, \\ L_{\rm pb} &= \Delta\lambda \pi \langle I_{\rm pb,\,norm}\rangle 4\pi \mathcal{R}_\odot^{\rm N^2} = 3.791 \times 10^{26}\ \text{W}. \end{aligned} \tag{4.43}$$

Here we used $\langle I_{\rm pb,\,norm}\rangle = \int_{90\ \rm nm}^{4\ \mu\rm m} B_\lambda d\lambda/\Delta\lambda = 5.074 \times 10^{12}\ \ \text{W m}^{-3}$. The percent difference between this quasi-bolometric passband and true bolometric values is ~1%.

Now let us do the same for the photon-counting devices. The equations in Table 4.2 simplify to:

$$\begin{aligned} F_{\rm pb}^{\rm pc} &= \mathcal{P}_{\rm int}^{\rm pc} \langle I_{\rm pb,\,norm}^{\rm pc}\rangle \pi \mathcal{R}_\odot^{\rm N^2}/D^2 = 5.911 \times 10^{21}\ \text{photons s}^{-1}\,\text{m}^{-2}, \\ L_{\rm pb}^{\rm pc} &= \mathcal{P}_{\rm int}^{\rm pc} \pi \langle I_{\rm pb,\,norm}^{\rm pc}\rangle 4\pi \mathcal{R}_\odot^{\rm N^2} = 1.662 \times 10^{45}\ \text{photons s}^{-1}, \end{aligned} \tag{4.44}$$

where $\mathcal{P}_{\rm int}^{\rm pc} = \int_\lambda \lambda \mathcal{P}(\lambda) d\lambda/hc = 4.025 \times 10^{13}\ \ \text{m}\ \ \text{J}^{-1}$ and $\langle I_{\rm pb,\,norm}^{\rm pc}\rangle = 2.161 \times 10^{12}$ photons $\text{s}^{-1}\ \ \text{m}^{-3}$. The percent difference now is a whopping 7%. This is not surprising, though: the larger discrepancy comes from weighting the intensity integral by λ, so any photons redward of the $4\ \mu$m cutoff are additionally weighted. In other words, the flux in energy units diminishes as λ^4 at long wavelengths (the so-called Rayleigh–Jeans regime), but only as λ^3 in photon counts. If we extended the red cutoff to $8\ \mu$m, the percent difference in photon counts would dip below 1%.

Now let us "turn on" limb darkening. We will model $I(\mu)$ with a logarithmic limb-darkening law (see Table 4.1):

$$\frac{I(\mu)}{I_0} = 1 - x_\lambda(1 - \mu) - y_\lambda\, \mu \ln \mu, \tag{4.45}$$

where x_λ and y_λ are passband-dependent limb-darkening coefficients. Their values depend on atmospheric parameters (effective temperature, surface gravity, heavy metal abundance, etc). For the Sun, the values are:

$$x_\lambda = 0.6923 \quad \text{and} \quad y_\lambda = 0.1628. \tag{4.46}$$

These were obtained by fitting $\mathcal{L}(\mu)$ to $I_\lambda(\mu)/I_{\lambda,\,\mathrm{norm}}$ values derived from Castelli & Kurucz (2004) model atmospheres within PHOEBE. For illustration purposes, let us use Equation (4.39) without correcting the Planckian intensity to demonstrate why this is a problem.

With limb darkening in effect, the $\mathcal{L}(\mu)$ integral is no longer unity; logarithmic $\mathcal{L}(\mu)$ is easily integrable:

$$\int_0^1 \mu\mathcal{L}(\mu)d\mu = \int_0^1 \mu[1 - x_\lambda(1 - \mu) - y_\lambda\mu \ln \mu] = 1 - \frac{1}{3}x_\lambda + \frac{2}{9}y_\lambda. \tag{4.47}$$

For the values given in Equation (4.46), this expression evaluates to $\mathcal{L}_{\mathrm{int}} = 0.8054$. Using energy-weighted equations in Table 4.2 yields:

$$\begin{aligned} L_{\mathrm{pb}} &= \Delta\lambda \int_0^{4\pi R^2} 2\pi\langle I_{\mathrm{pb,\,norm}}\rangle\mathcal{L}_{\mathrm{int}}dA = 3.053 \times 10^{26}\ \mathrm{W}, \\ F_{\mathrm{pb}} &= \frac{\Delta\lambda}{D^2}\int_0^{2\pi R^2} \langle I_{\mathrm{pb,\,norm}}\rangle\mu\mathcal{L}(\mu)dA = 1085.323\ \mathrm{Wm^{-2}}. \end{aligned} \tag{4.48}$$

These values should raise a red flag: the intensities $\langle I_{\mathrm{pb,\,norm}}\rangle$ and limb-darkening coefficients were computed for $T_{\mathrm{eff}} = 5772$ K, but the flux we obtain corresponds to 5468 K. The difference corresponds precisely to $\mathcal{L}_{\mathrm{int}}$ (its 1/4 power, to be exact) in Equation (4.39). Thus, if we replace $I_{\lambda,\,\mathrm{norm}}$ with $I_{\lambda,\,\mathrm{norm}}/\mathcal{L}_{\mathrm{int}}$, we reproduce the expected values:

$$\begin{aligned} L_{\mathrm{pb}} &= \Delta\lambda \int_0^{4\pi R^2} 2\pi\langle I_{\mathrm{pb,\,norm}}\rangle dA = 3.791 \times 10^{26}\ \mathrm{W}, \\ F_{\mathrm{pb}} &= \frac{\Delta\lambda}{D^2}\int_0^{2\pi R^2} \frac{\langle I_{\mathrm{pb,\,norm}}\rangle}{\mathcal{L}_{\mathrm{int}}}\mu\mathcal{L}(\mu)dA = 1347.539\ \mathrm{Wm^{-2}}. \end{aligned} \tag{4.49}$$

The same procedure applies to the photon-weighted values, yielding:

$$\begin{aligned} F_{\mathrm{pb}}^{\mathrm{pc}} &= 5.909 \times 10^{21}\ \mathrm{photons\ s^{-1}\ m^{-2}}, \\ L_{\mathrm{pb}}^{\mathrm{pc}} &= 1.662 \times 10^{45}\ \mathrm{photons\ s^{-1}}. \end{aligned} \tag{4.50}$$

Switching now to model atmospheres, there is one important caveat to be aware of: I_μ already "contains" the information about limb darkening by way of the numerically estimated source function $\mathcal{S}_\lambda$. Thus, if limb darkening is turned off, we *expect* that both the luminosity and flux are overestimated. This does not technically make the solution inconsistent: if there were no limb darkening, those would be true luminosity and flux values; the culprit for an apparently inconsistent result is the unphysical assumption that we make by switching off limb darkening.

Beyond that consideration, all other principles used in blackbody computation remain the same. That is why we summarize the numerical values in Table 4.3 and end the discussion by stressing yet again that, although relatively straightforward, the computation of radiative properties is wrought with intricacies that are further

Table 4.3. Luminosities, Fluxes, and Scaled Fluxes in Energy- and Photon-weighted Mode

LD Model	$L_{\rm pb}$ (W)	$F_{\rm pb}$ (W m^{-2})	$\xi F_{\rm pb}$ (W m^{-2})	$L_{\rm pb}^{\rm pc}$ (photons s^{-1})	$F_{\rm pb}^{\rm pc}$ (photons s^{-1} m^{-2})	$\xi F_{\rm pb}^{\rm pc}$ (W m^{-2})
None	4.714×10^{26}	1676.190	1347.909	2.031×10^{45}	7.224×10^{21}	1347.910
Logarithmic	3.797×10^{26}	1349.649	1347.538	1.636×10^{45}	5.816×10^{21}	1347.538
Interpolated	3.842×10^{26}	1365.777	1347.507	1.734×10^{45}	6.165×10^{21}	1347.560

Note. The values pertain to Castelli & Kurucz (2004) model atmospheres and solar atmospheric parameters.

complicated by the presence of another body in a system—a topic of the following section.

References

Brown, A. G. A., Vallenari, A., Prusti, T., et al. 2016, A&A, 595, A2
Castelli, F., & Kurucz, R. L. 2004, arXiv:astro-ph/0405087
Claret, A. 2000, A&A, 363, 1081
Claret, A., & Bloemen, S. 2011, A&A, 529, A75
Claret, A., & Hauschildt, P. H. 2003, A&A, 412, 241
Diaz-Cordoves, J., & Gimenez, A. 1992, A&A, 259, 227
Espinoza, N., & Jordán, A. 2016, MNRAS, 457, 3573
Gray, D. F. 2008, The Observation and Analysis of Stellar Photospheres (Cambridge: Cambridge Univ. Press)
Hauschildt, P. H., Allard, F., & Baron, E. 1999, ApJ, 512, 377
Kipping, D. M. 2013, MNRAS, 435, 2152
Klinglesmith, D. A., & Sobieski, S. 1970, AJ, 75, 175
Kopal, Z. 1950, HarCi, 454, 1
Kurucz, R. L. 2013, ATLAS12: Opacity Sampling Model Atmosphere Program, Astrophysics Source Code Library, ascl:1303.024
Lester, J. B., & Neilson, H. R. 2008, A&A, 491, 633
Neilson, H. R., & Lester, J. B. 2013, A&A, 556, A86
Orosz, J. A., & Hauschildt, P. H. 2000, A&A, 364, 265
Prša, A., Conroy, K. E., Horvat, M., et al. 2016a, ApJS, 227, 29
Prša, A., Harmanec, P., Torres, G., et al. 2016b, AJ, 152, 41
Schwarzschild, K. 1906, NWGot, 195, 41

Chapter 5

Radiative Properties of Binary Stars

The distribution of brightness of the surface of tidally and rotationally distorted stars is anything but trivial. A range of phenomena need to be accounted for in order to get a reasonable model for binary stars. In this chapter we review the most important contributions to surface brightness variation: gravity darkening, reflection, spots, Doppler boosting, and interstellar/atmospheric extinction.

5.1 Gravity Darkening

The distribution of intensity over the stellar disk depends on the dominant energy transfer mechanism in a stellar envelope. The mechanism is primarily determined by the effective temperature of the star: for $T_{\text{eff}} > 8000$ K the envelope is in a predominantly radiative equilibrium, whereas for $T_{\text{eff}} < 5000$ K the envelope is in a predominantly convective equilibrium (Claret 1999); in between, both mechanisms significantly contribute to the energy transfer.

For a rotating star in a strict radiative equilibrium, a pioneering work by von Zeipel (1924) has shown that the flux distribution over the surface is proportional to the effective gravity:

$$F_\lambda = -\frac{16\sigma T^3}{3\bar{\kappa}\rho}\frac{dT}{d\Omega}g^\beta, \tag{5.1}$$

where $\sigma = 5.67 \times 10^{-8}$ W m^{-2} K^{-4} is the Stefan–Boltzmann constant, T is the local surface temperature, $\bar{\kappa}$ is the Rosseland mean opacity, ρ is the density of the gas, g is the local gravity acceleration, and β is the gravity-darkening coefficient, for which von Zeipel (1924) rigorously showed that $\beta = 1$ for strictly radiative envelopes. Equation (5.1) implies that the equatorial regions are darker than the polar ones, hence the name "gravity darkening." The local temperature may also be coupled to the gravitational acceleration:

$$T \propto F^{1/4} \Rightarrow T \propto g^{\beta/4}. \tag{5.2}$$

doi:10.1088/978-0-7503-1287-5ch5

The generalized formalism of computing the value of the gravity-darkening coefficient β for radiative envelopes was presented later by Kippenhahn (1977).

Prediction for the gravity-darkening exponent β of purely convective envelopes was first done by Lucy (1967), who used convective stellar envelope models to compute the average value $\beta = 0.32$ for solar-type stars. The driving idea of Lucy's study was that, within the fully convective zone, the gradient of specific entropy is 0. His predictions for the gravity-darkening coefficient were later confirmed observationally by Rafert & Twigg (1980), who obtained the average value $\beta = 0.31$ for the sample of 28 stars.

Putting the two results for radiative and convective equilibrium together, gravity darkening is given by:

$$F = F_{\text{pole}}\left(\frac{g}{g_{\text{pole}}}\right)^{\beta}, \quad \beta = \begin{cases} 1.00 & \text{for radiative envelopes} \\ 0.32 & \text{for convective envelopes.} \end{cases} \tag{5.3}$$

Factor

$$\mathcal{G} = \frac{F}{F_{\text{pole}}} = \left(\frac{g}{g_{\text{pole}}}\right)^{\beta} = T_{\text{pole}}\left(\frac{g}{g_{\text{pole}}}\right)^{\beta/4} \tag{5.4}$$

is called the gravity-darkening correction.

Two later studies by Alencar & Vaz (1997) and Alencar et al. (1999) followed Lucy's approach of calculating gravity darkening for convective stellar atmospheres; they used nonilluminated and illuminated Uppsala model atmospheres (Gustafsson 1983), respectively, deriving an empirical formula for $\beta(T_{\text{eff}})$, depicted in Figure 5.1 (left panel). The results indicate agreement with Lucy's estimate only at temperatures $\sim$ 6500 K.

Since the gravity-darkening effect depends on the circumstances in stellar envelopes, it is related not only to the atmospheric parameters (T_{eff} and $\log g$) but also to the internal stellar structure and rotation properties. This was pointed out by Claret (1999), who introduced the computation of gravity-darkening coefficients into the stellar evolution code. Statistical studies based on this new formalism (Claret 2000, Figure 1 therein) yielded a more accurate theoretical prediction for stellar envelopes where both radiation and convection contribute significantly to energy transfer.

The gravity-darkening correction given by Equation (5.4) is *bolometric*. Thus, to apply gravity darkening appropriately to passband data, a bolometric correction needs to be made. Such a correction has been first proposed by Martynov (1973) and later expanded by Bloemen et al. (2011) to the following form:

$$\beta(\lambda;\, T_{\text{eff}}, \log g, \ldots) = \left(\frac{\partial \log I(\lambda)}{\partial \log g}\right)_{T_{\text{eff}}} + \left(\frac{\partial \log T_{\text{eff}}}{\partial \log g}\right)\left(\frac{\partial \log I(\lambda)}{\partial\, \text{T}_{\text{eff}}}\right)_{\log g}. \tag{5.5}$$

The right panel of Figure 5.1 depicts the dependence of β on temperature for the *Kepler* passband (Claret & Bloemen 2011).

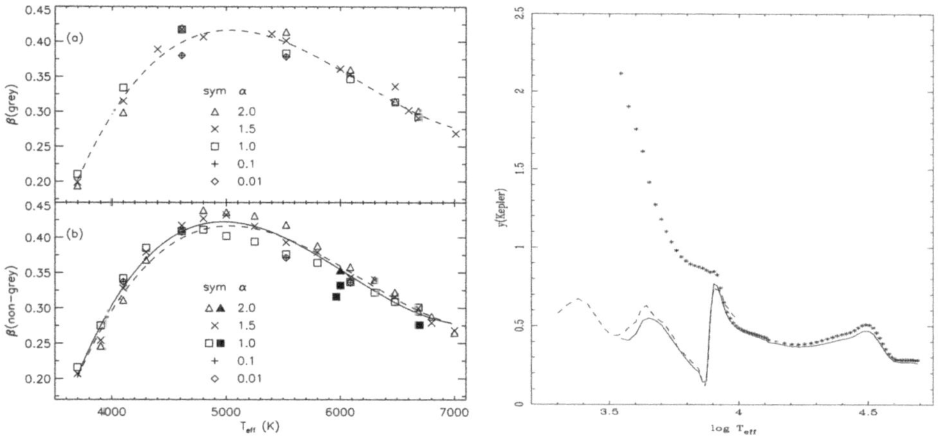

Figure 5.1. Gravity-darkening coefficient β dependence on temperature for convective envelopes. Left: results for gray (i.e., continuum opacity) and nongray (i.e., line-blanketed opacity) Uppsala model atmospheres computed by Alencar & Vaz (1997). Symbols correspond to different convection efficiency coefficients $\alpha = 1/H_p$. Right: dependence of the gravity-darkening coefficient on temperature for the *Kepler* passband (Claret & Bloemen 2011). The solid line corresponds to Equation (5.5) fit to Kurucz model atmospheres, the dashed line is a fit to Phoenix model atmospheres, and asterisks correspond to Martynov's simplified equation fit to Kurucz model atmospheres. Credit: Alencar & Vaz, A&A, 326, 257, 1997, reproduced with permission © ESO. Credit: Claret and Bloemen, A&A, 529, 529, A75, 2011, reproduced with permission © ESO.

Espinosa Lara & Rieutord (2011) also pointed out the failure of von Zeipel's theorem for rapidly rotating stars. They proposed a model that describes latitudinal variation of gravity darkening as a consequence of rotation. In their subsequent paper, Espinosa Lara & Rieutord (2012) further developed the formalism for gravity darkening in binary stars and advocated for the abandonment of a single-parameter description and replacing it with the following expression:

$$f(x, y, z) = f_0 \exp\left(\int_0^{r_{\rm point}} \frac{2(1+q)}{g_{\rm eff}^2(x, y, z)} g_{\rm eff}(x, 0,0)dx\right), \tag{5.6}$$

where $f(x, y, z)$ is surface flux at point (x, y, z), $f_0 = \lim_{r\to 0} f = L/4\pi GM$, $\mathbf{g}_{\rm eff} = -\nabla\psi$ is the effective gravitational acceleration, and $g_{\rm eff}$ is the magnitude of $\mathbf{g}_{\rm eff}$. This integral can be solved numerically to obtain a relationship between the flux and gravitational acceleration.

5.2 Reflection Effect

When components in a binary system are close, i.e., when their radii are ~ 15%–20% of their separation or more (Wilson 1990), surface irradiation of one component by the other becomes important. A part of the incident flux heats the surface of the irradiated star, increasing the temperature, and the other part of the incident flux is reflected off the irradiated star. This effect is called the reflection effect.

Before we develop the formalism to account for the reflection effect, we should identify the cases where reflection contributes significantly to the overall flux of the irradiated component. There are indeed quite a number of such cases, ranging from stars similar in size and effective temperature (Hilditch et al. 1996), to cataclysmic and pre-cataclysmic binaries (Pigulski & Michalska 2002), to rapidly rotating Algols (Olson & Etzel 1994), to elliptical ellipsoidal variables known as heartbeat stars (Welsh et al. 2011).

The rigorous treatment of the reflection effect in binary stars was first developed by Wilson (1990); before that work only approximate solutions existed—see a thorough review by Vaz (1985) on this topic.

Consider a close binary as depicted in Figure 5.2. To properly account for the heating of a given surface element dA_1 on the primary star, we need to sum all contributions to irradiation over all surface elements dA_2 on the secondary star. Following the energy conservation principle, recall that heating is a bolometric process, whereas radiation is wavelength dependent, described by the spectral energy distribution (SED) function (blackbody, model atmospheres, ...). Bolometric emergent intensity I_2^{bol} emitted from surface element dA_2 into the direction of dA_1 may be expressed as the limb-darkened normal bolometric emergent intensity $I_{2\perp}^{\mathrm{bol}}$:

$$I_2^{\mathrm{bol}} = I_{2\perp}^{\mathrm{bol}} \mathcal{L}_2^{\mathrm{bol}}(\mu_2), \tag{5.7}$$

where $\mu_2 = \cos\psi_2$ is the cosine of the angle between the surface normal to dA_2 and the direction $\boldsymbol{s}$ that connects the two surface elements. The incident intrinsic flux from dA_2 to dA_1 can then be written per Equation (4.15) as:

$$dF_{2\to1}^{\mathrm{bol}} = \frac{\mu_1\mu_2}{s^2} I_{2\perp}^{\mathrm{bol}} \mathcal{L}_2^{\mathrm{bol}}(\mu_2) dA_2, \tag{5.8}$$

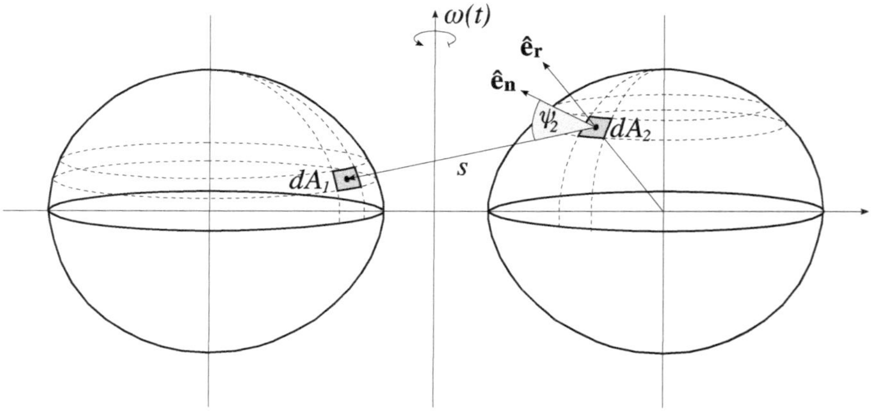

Figure 5.2. Irradiation of the primary star (left) by the secondary star (right). To properly account for the heating of the surface element dA_1 on the primary star, contributions to irradiation of all surface elements dA_2 on the secondary star must be summed up. Angle ψ_2 is the angle between the surface normal $\hat{\mathbf{e}}_\mathbf{n}$ and the direction $\boldsymbol{s}$ connecting the two surface elements. Its counterpart ψ_1 is not depicted for decluttering purposes. Angles γ_1 and γ_2 between $\hat{\mathbf{e}}_\mathbf{n}$ and $\hat{\mathbf{e}}_\mathbf{r}$ are also not depicted.

where $\mu_1 = \cos\psi_1$ is the cosine of the incident angle between the surface normal to dA_1 and the direction $-\boldsymbol{s}$.

We now introduce a reflection factor $\mathcal{R}$ as the ratio of the total (intrinsic + reflected) flux to intrinsic flux. If there is no reflection, $\mathcal{R} \equiv 1$. The total incident flux from dA_2 to dA_1 can then be computed by multiplying the intrinsic incident flux, given by Equation (5.8), by $\mathcal{R}_2$:

$$F_{2\to1}^{\rm bol} = \int_{\partial V} \mathcal{R}_2 \mathcal{V}(\boldsymbol{r}) \frac{\mu_1\mu_2}{s^2} I_{2\perp}^{\rm bol} \mathcal{L}_2^{\rm bol}(\mu_2)\, dA_2, \tag{5.9}$$

where we used the visibility function $\mathcal{V}(\boldsymbol{r})$ defined by Equation (4.13), this time pertaining to the visibility of surface elements along the line s.

A fraction of the total incident flux will be used for heating the irradiated surface, and a fraction will be reflected. The two fractions need to sum up to 1, and they are determined by the bolometric *albedo* of the star, $\mathcal{A}$. It is the ratio of the reradiated flux to irradiated flux. Thus, the fraction of reflected flux is given by:

$$F_{1,\rm refl}^{\rm bol} = \mathcal{A}_1 \int_{\partial V} \mathcal{R}_2 \mathcal{V}(\boldsymbol{r}) \frac{\mu_1\mu_2}{s^2} I_{2\perp}^{\rm bol} \mathcal{L}_2^{\rm bol}(\mu_2)\, dA_2. \tag{5.10}$$

The intrinsic bolometric normal emergent intensity $I_{1\perp}^{\rm bol}$ emitted from the surface element dA_1 is computed from the polar emergent intensity $I_{\rm polar}^{\rm bol}$ corrected for gravity darkening via Equation (5.4). Note that Equation (5.4) relates fluxes and not intensities, but as these are bolometric intensities (integrated over all wavelengths), it means that $F_{\rm bol} = \pi I_{\rm bol}$ per Stefan–Boltzmann's law, irrespective of the actual distribution of $I(\lambda)$. In other words, *all* models for $I(\lambda)$ (blackbody, Kurucz, …), when integrated over all λ, must yield $\sigma T_{\rm eff}^4/\pi$.

The intrinsic bolometric flux $F_{1,\rm intr}^{\rm bol}$ is computed per Equation (4.5):

$$F_{1,\rm intr}^{\rm bol} = 2\pi I_{1\perp}^{\rm bol} \int \mathcal{L}_1^{\rm bol}(\mu_1)\, \mu_1\, d\mu_1. \tag{5.11}$$

The ratio $F_{1,\rm refl}^{\rm bol}/F_{1,\rm intr}^{\rm bol}$ is easily related to the reflection factor $\mathcal{R}_1$:

$$\mathcal{R}_1 = \frac{F_{1,\rm intr}^{\rm bol} + F_{1,\rm refl}^{\rm bol}}{F_{1,\rm intr}^{\rm bol}} = 1 + \frac{F_{1,\rm refl}^{\rm bol}}{F_{1,\rm intr}^{\rm bol}}. \tag{5.12}$$

Similarly, for the other star,

$$\mathcal{R}_2 = \frac{F_{2,\rm intr}^{\rm bol} + F_{2,\rm refl}^{\rm bol}}{F_{2,\rm intr}^{\rm bol}} = 1 + \frac{F_{2,\rm refl}^{\rm bol}}{F_{2,\rm intr}^{\rm bol}}. \tag{5.13}$$

Equations (5.9), (5.10), and (5.12) thus form a recursive set to compute $\mathcal{R}_1$, and their counterparts for the secondary star form a recursive set to compute $\mathcal{R}_2$. In practice, this means that we start from some initial values for $\mathcal{R}_1$ and $\mathcal{R}_2$, say, unity, and then iterate the recursion set until we reach a desired accuracy.

Once the converged values of $\mathcal{R}_1$ and $\mathcal{R}_2$ are computed, we can calculate a new effective temperature of the surface element that encompasses the radiation of both intrinsic and reflected energy:

$$T_{\mathrm{eff1,new}} = \mathcal{R}_1^{1/4} T_{\mathrm{eff1,old}}, \quad \text{and} \quad T_{\mathrm{eff2,new}} = \mathcal{R}_2^{1/4} T_{\mathrm{eff2,old}}. \tag{5.14}$$

We can do this only because all quantities that partake in reflection are bolometric.

The above procedure may be repeated iteratively any number of times to accommodate for multiple reflections: the total emergent flux from the primary star now heats the secondary star by an amount greater than what the intrinsic emergent flux by itself would cause.

A keen eye will catch that this description neglects the fraction of the incident flux that is not reflected. This may or may not be a problem: if we assume that the absorbed energy is redistributed across the entire star, the effect will not be significant. However, if we assume that the absorbed energy is localized to the immediate vicinity of the irradiated area, then we expect an additional local temperature increase. This was first worked out by Prša et al. (2016). The basic difference in the treatment is the energy balance equation (see Figure 5.3):

$$F_{\mathrm{in}}(\boldsymbol{r}) = \hat{\mathcal{L}}_{\mathrm{LD}} F_0(\boldsymbol{r}) + \hat{\mathcal{L}}_{\mathrm{L}}(\rho F_{\mathrm{in}})(\boldsymbol{r}), \tag{5.15}$$

$$F_{\mathrm{out}}(\boldsymbol{r}) = F_0(\boldsymbol{r}) + \rho(\boldsymbol{r}) F_{\mathrm{in}}(\boldsymbol{r}), \tag{5.16}$$

where $\hat{\mathcal{L}}$ is the *radiosity* operator that describes the physical process of heating and reflection/scattering. The two simplest processes are Lambertian, described by $\hat{\mathcal{L}}_{\mathrm{L}} = \mathcal{R}/\pi$, and limb-darkened distribution, described by $\hat{\mathcal{L}}_{\mathrm{LD}} = \mathcal{R} \circ \mathcal{L}$. An expanded reflection theory has been developed by Horvat et al. (2018), which includes different energy redistribution modes: local, latitudinal, and global (Figure 5.4). Their approach further generalizes the treatment of reflection, but it comes at a significantly higher computational cost.

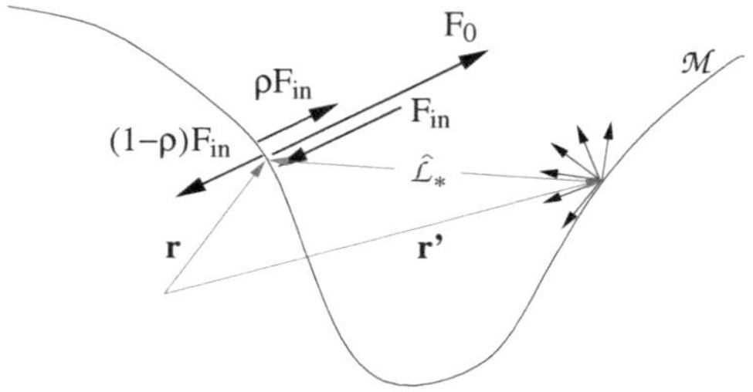

Figure 5.3. Schematic representing the energy balance given the total irradiation of the element at $\boldsymbol{r}$. F_{in} is the total incident bolometric flux, ρF_{in} is the fraction of F_{in} that is reflected, $(1-\rho)F_{\mathrm{in}}$ is the fraction of F_{in} that is absorbed, and $\hat{\mathcal{L}}_*$ is the radiosity operator that describes the physical process of heating and reflection/scattering. The total emergent flux is denoted with F_0.

5.3 Stellar Spots

The discovery of chromospheric activity of the Sun dates back to prehistoric times (ancient China and ancient Greece), but it was not until Galileo & Harriot observed them in 1610 and until Fabricius published his observations in 1611 that the world truly embraced the concept of sunspots, challenging the perfect, unchanging nature of the universe. Proposing spots on other stars followed shortly, by Boulliau (1667), but we had to wait for actual observations of spots until Kron (1947), who observed eclipsing binary AR Lac and invoked starspots to explain aperiodic out-of-eclipse variability. Over the next several decades we learned that starspots are connected to stellar magnetic fields (Spruit & Weiss 1986), so we expect most (if not all) convective envelope stars to feature spots. Another type of spot is related to mass transfer and accretion in cataclysmic variables (Warner 2003), which are hotter and brighter than the underlying photosphere. A comprehensive review of our current understanding of starspots is given by Strassmeier (2009).

Modern observations of starspots abound; Figure 5.5 depicts a light curve of KIC 2438502, an eclipsing binary with an 8.36-day period that features an RS CVn-type component that rotates with a period of ~8.23 days. Most out-of-eclipse variation in this light curve originates in chromospheric activity, which is not periodic; hence, it results in the smearing of the phased light curve. Starspot evolution of two persistent active longitudes for this object has been reported and studied by Roettenbacher et al. (2016).

In terms of modeling, spots may be described by four parameters:

SPOT COLATITUDE θ_{sp}: the colatitude of the spot on the stellar surface, described in the same coordinate system as the star itself.

SPOT LONGITUDE ϕ_{sp}: the longitude of the spot on the stellar surface, described in the same coordinate system as the star itself.

SPOT ANGULAR RADIUS ϱ_{sp}: angular size of the spot, usually given in radians.

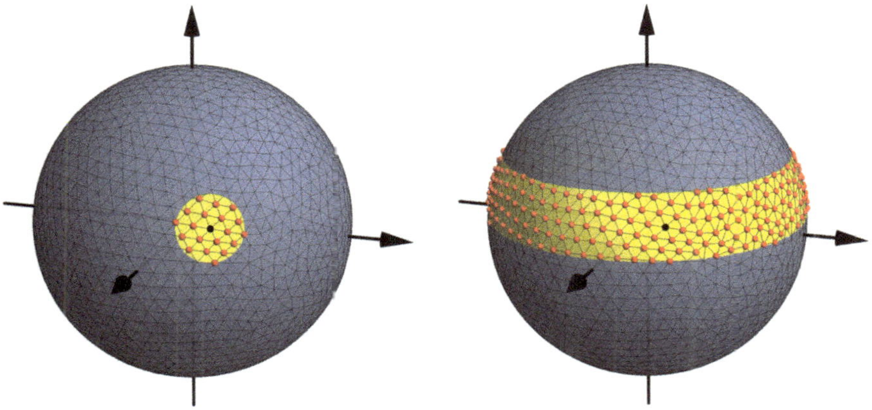

Figure 5.4. Local (left) and latitudinal (right) redistribution of energy being absorbed at the surface element depicted in black and emitted from the surface areas depicted in red. The regions depicted in yellow represent the area affected by redistribution. Adopted from Horvat et al. (2018).

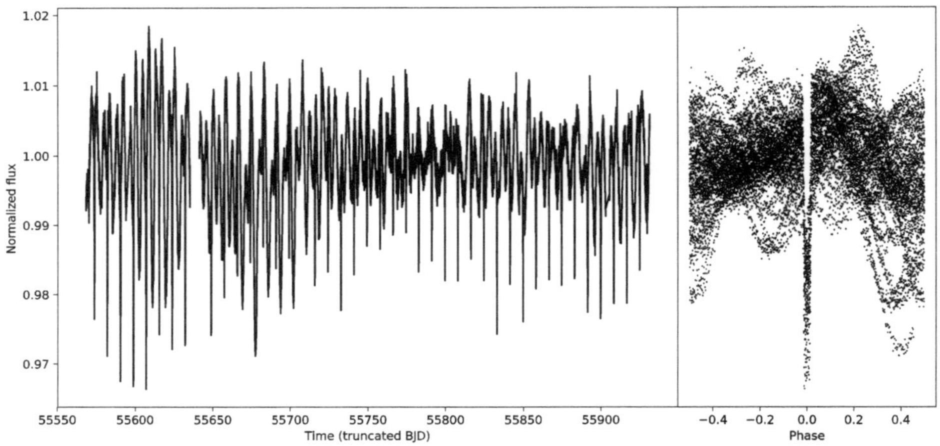

Figure 5.5. *Kepler* light curve of KIC 2438502, an RS CVn-type eclipsing binary star. The dominant variations are a consequence of chromospheric activity. The time series (left) and phased light curve (right) show that spot activity is neither periodic nor commensurate with the orbital period of the binary.

SPOT TEMPERATURE FACTOR $\tau_{\rm sp}$: the ratio between the temperature of the spot and the local temperature of the underlying photosphere. Values $\tau_{\rm sp} > 1$ correspond to hot spots, while values $\tau_{\rm sp} < 1$ correspond to cool spots.

The key assumption when modeling the area covered by a spot is that its local temperature is modified by the temperature factor and the area then radiates as if it were a normal photosphere at that modified temperature.

We can justly question the adequacy of this simplistic description: to properly model spots, advanced and reliable methods of data acquisition (polarimetry, Doppler tomography) are required (see chapter 2). It is *not* straightforward to use the data in the sophisticated treatment of spots (e.g., nonuniform spot brightness, magnetohydrodynamic effects, etc.), especially as introducing spots without clear support by the data may easily reproduce *any* physical effect—the degeneracy is typically severe. Even such a simplified treatment, along with any temporal evolution of spot parameters, is thus still ahead of the majority of our current observational capabilities.

5.4 Doppler Boosting

The amount of light that reaches the observer is affected by the kinematic properties of the radiating body. There are three important velocity-induced changes to the received flux: (1) Doppler shift of the spectrum causes the passband-weighted integral to change (Equations (4.27) and (4.28)); (2) Doppler shift of the frequency causes the arrival rate of the photons to change because of time dilation; and (3) relativistic beaming, where radiation is no longer isotropic but becomes direction-dependent because of light aberration (Rybicki & Lightman 1979, chap. 4.8). Loeb & Gaudi (2003) showed that the boosting signal in exoplanet-hosting stars is larger than variability due to reflected light from the planet; Zucker et al. (2007) performed a similar study for binary stars and showed that ellipsoidal variability, reflection, and boosting are same-order-of-magnitude effects in binaries with a ~ 10-day orbital period, while boosting dominates in binaries with a ~ 100-day orbital period. Studies by van Kerkwijk et al. (2010) and Bloemen et al. (2011, 2012) apply this correction to *Kepler* objects KOI-74, KOI-81, and KPD 1946+4340 and demonstrate that the contributions of boosting are indeed crucial to model the light curves correctly.

The combined Doppler boosting signal can be written as:

$$I_\lambda = I_{0,\lambda}\left(1 - \mathcal{B}(\lambda)\frac{v_r}{c}\right), \tag{5.17}$$

where I_λ is the boosted passband intensity, $I_{0,\lambda}$ is the initial passband intensity, v_r is the radial velocity, c is the speed of light, and $\mathcal{B}(\lambda)$ is the boosting index,

$$\mathcal{B}(\lambda) = 5 + \alpha \equiv 5 + \frac{d \ln I}{d \ln \lambda}, \tag{5.18}$$

where $\alpha \equiv d(\ln I)/d(\ln \lambda)$ is the spectral index, and 5 comes from the Lorenz invariance of $I\lambda^5$ (Loeb & Gaudi 2003). Figure 5.6 depicts the boosting index for the broad Johnson V passband, for a series of $T_{\rm eff} = 6000$ K, $\log g = 4.0$, [M/H] = 0.0 spectra where μ ranges from 0.001 to 1.0. The dependence of the boosting index on wavelength is notable, but in the simplified approach we approximate the monochromatic boosting factor $\mathcal{B}(\lambda)$ with the passband-averaged value:

$$\mathcal{B}_{\rm pb} = \frac{\int_\lambda \mathcal{P}(\lambda)\mathcal{S}(\lambda)\mathcal{B}(\lambda)d\lambda}{\int_\lambda \mathcal{P}(\lambda)\mathcal{S}(\lambda)d\lambda} \tag{5.19}$$

(or its photon-weighted counterpart per Equation (4.28)). This is done by fitting a Legendre polynomial of the fifth order to the $\ln I_\mu(\ln \lambda)$ data; the order is determined by evaluating the rank of the coefficient matrix in the least-squares fit and seeing where the values become susceptible to numerical noise. The series was fit iteratively, as data points that lie below the -1σ threshold were discarded by sigma-clipping the data set. The iteration stopped when no further points are removed. Another benefit of Legendre polynomials is their analytical derivative; this analytic form is used to derive the average boosting index for every model atmosphere. Figure 5.6 also shows that the dependence of the boosting index on μ is significant, so the traditional approach of using integrated flux SEDs to estimate $\mathcal{B}_{\rm pb}$ is less precise than the treatment presented here.

Boosting is a significant effect, and it becomes a dominant effect over ellipsoidal variation and reflection for longer-period systems. Figure 5.7 shows a comparison of the amplitude of ellipsoidal variation (blue), reflection (green), and Doppler boosting (red) as a function of orbital period for an A0–K0 main-sequence pair in the Johnson V passband (Prša et al. 2016). Ellipsoidal variation and reflection dominate the short-period end, with boosting taking over at around 8 days. Figure 5.8 depicts a light curve with no effects computed (black), with only ellipsoidal

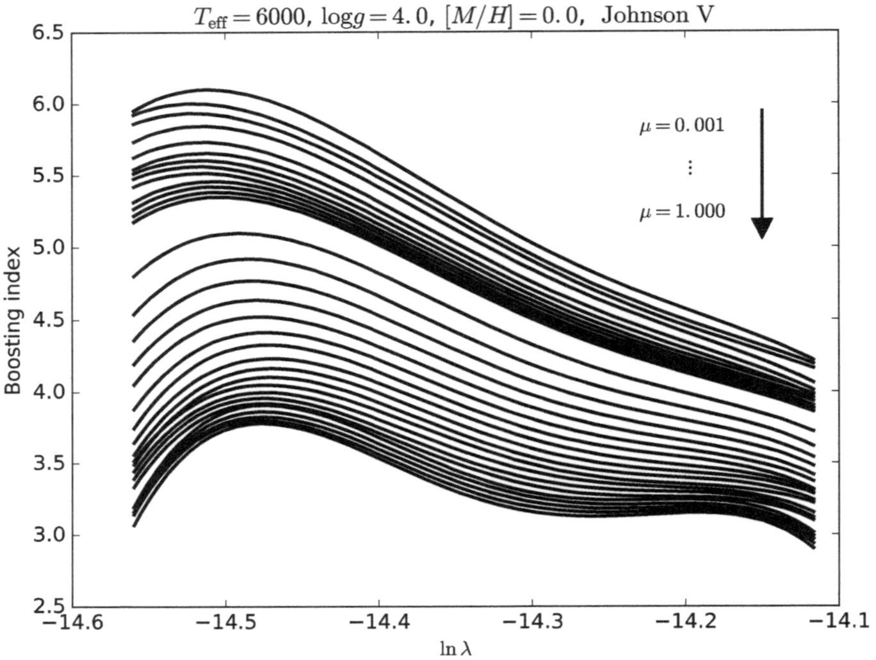

Figure 5.6. Boosting index $\mathcal{B}(\lambda)$ for the Johnson V broadband filter. The spectra correspond to $T_{\rm eff} = 6000$ K, $\log g = 4.0$, [M/H] = 0.0, and μ runs over 33 values ranging from 0.001 to 1.0. The dependence of the boosting index on wavelength is obvious, but at this time we approximate it with a weighted average across the passband (see Equation (5.19)). The increased curviness at larger μ is due to the larger variations and larger fluxes as we look toward the center of the stellar disk. Taken from Prša et al. (2016). © 2016. The American Astronomical Society.

variations (blue), with ellipsoidal variations and reflection (green), and with ellipsoidal variations, reflection, and boosting (red).

5.5 Interstellar and Atmospheric Extinction

Interstellar extinction is an extrinsic effect that causes a wavelength-dependent reduction of light that reaches the observer. It was first reported by Trumpler (1930) and connected with dust grains that form the interstellar medium. Extinction is also known as reddening because it affects the blue part of the spectrum more than the red, making objects appear redder than they are intrinsically (Draine 2003). It is usually described by the color excess:

$$E(B - V) = (B - V)_{\rm observed} - (B - V)_{\rm intrinsic}. \tag{5.20}$$

It provides a measure of reddening in terms of the $B - V$ color index change. The slope of the extinction curve in the optical part of the spectrum is given by:

$$R_V \equiv \frac{A_V}{A_B - A_V}, \tag{5.21}$$

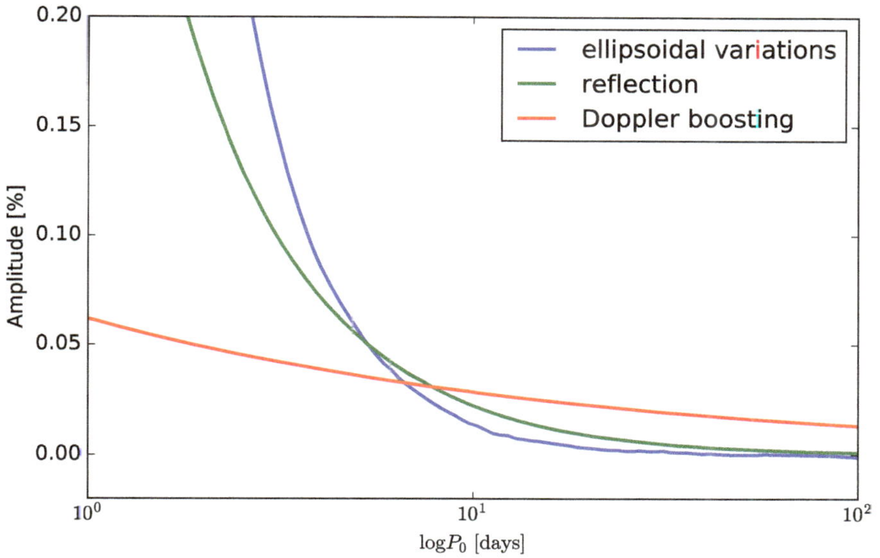

Figure 5.7. Amplitudes of ellipsoidal variation (blue), reflection (green), and boosting (red) for an A0–K0 main-sequence pair in the Johnson V passband as a function of orbital period. The semimajor axis is constrained to conserve the masses of components. For this particular system, ellipsoidal variations dominate the short-period end, reflection takes over at ~ 5.3 days, and boosting takes over at ~ 7.7 days. Taken from Prša et al. (2016). © 2016. The American Astronomical Society.

where A_B and A_V are extinctions (in mag) in the B and V passbands, respectively. The value of R_V depends on the grain size. For example, Rayleigh scattering similar to that in our atmosphere ($A(\lambda) \propto \lambda^{-4}$) would produce a slope of $R_V \sim 1.2$; extinction in our Galaxy ranges between ~2 and ~6, with an average around $R_V \sim 3.1$. The larger the grain size, the larger the R_V; in the limit of very large grain sizes, $R_V \to \infty$ (Draine 2003). This regime is called gray extinction, i.e., extinction does not depend on the wavelength.

The most frequently used model to describe wavelength dependence of interstellar extinction is that of Cardelli et al. (1989). It is an empirical formula with the assumed extinction slope $R_V = 3.1$. The formula is split into three wavelength regions:

Region	λ-interval	$x = (1/\lambda)$-interval
UV	$170\ \text{nm} \leqslant \lambda < 303\ \text{nm}$	$3.3\ \mu\text{m}^{-1} \leqslant x < 5.9\ \mu\text{m}^{-1}$
Optical/NIR	$303\ \text{nm} \leqslant \lambda < 910\ \text{nm}$	$1.1\ \mu\text{m}^{-1} \leqslant x < 3.3\ \mu\text{m}^{-1}$
IR	$910\ \text{nm} \leqslant \lambda < 3330\ \text{nm}$	$0.3\ \mu\text{m}^{-1} \leqslant x < 1.1\ \mu\text{m}^{-1}$

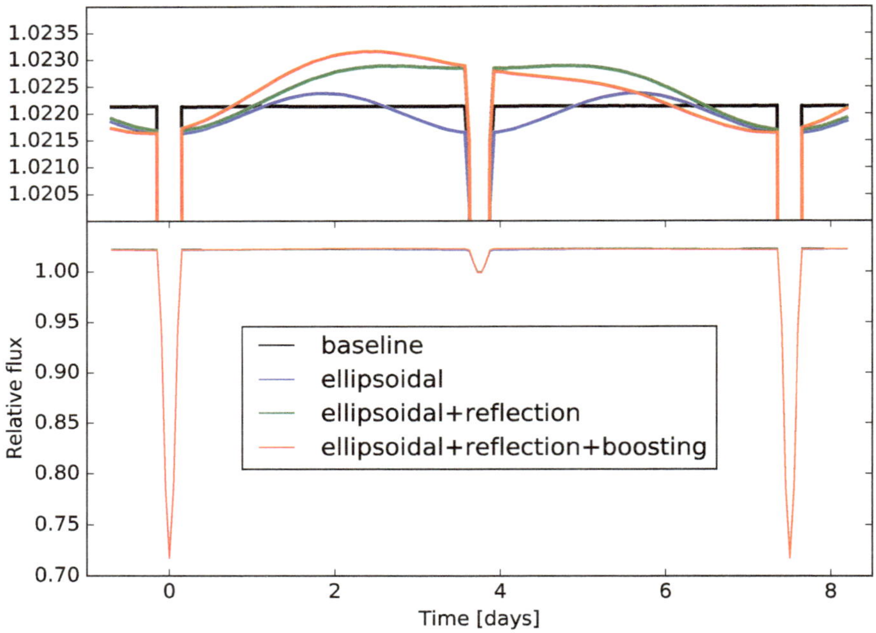

Figure 5.8. Light curve of an A0–K0 main-sequence pair in a 7.5-day orbit. The top panel is the zoomed-in version of the bottom panel. Four models are being plotted: the baseline model (black), which assumes spherical stars; the ellipsoidal variation model (blue), which includes surface deformation due to tides and rotation; the ellipsoidal variation model with reflection (green), which introduces reflection; and the complete model (red), which accounts for Doppler boosting on top of other effects. While the effects are small in amplitude as evident from the bottom panel, high-precision (mmag or better) photometry will routinely demonstrate these effects. Adopted from Prša et al. 2016 © 2016. The American Astronomical Society.

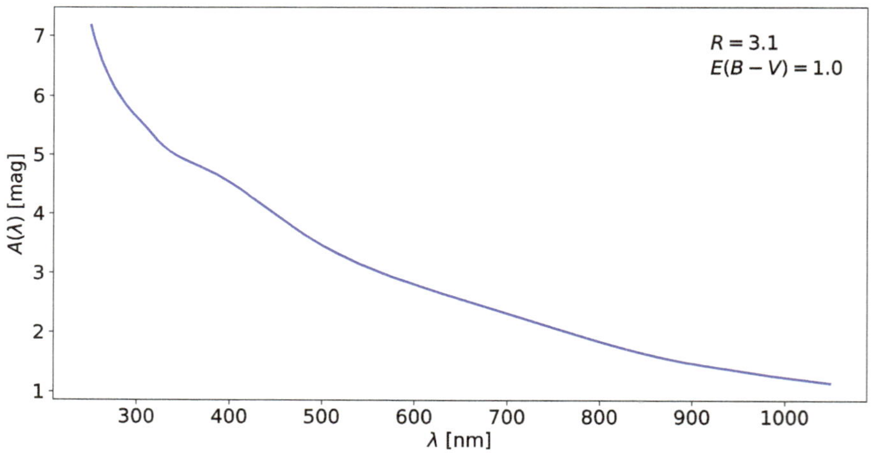

Figure 5.9. Reddening model of Cardelli et al. (1989). Visual extinction slope $R_V = 3.1$ and color excess $E(B - V) = 1.0$ are assumed.

The amount of interstellar extinction at a given wavelength (see Figure 5.9) is described by the interpolation formula:

$$\left\langle \frac{A(\lambda)}{A_V} \right\rangle = a(x) + b(x)/\mathcal{R}_V, \tag{5.22}$$

where $A(\lambda)$ is extinction in mag at the given wavelength, expressed relative to the visual extinction A_V; $x = 1/\lambda$ is given in μm^{-1}. Coefficients $a(x)$ and $b(x)$ are computed by the following expressions:

Region	
UV	$a(x) = 1.752 - 0.316x - 0.104/[(x - 4.67)^2 + 0.341] - 0.04473(x - 5.9)^2 - 0.009779(x - 5.9)^3$
	$b(x) = -3.090 + 1.825x + 1.206/[(x - 4.62)^2 + 0.263] + 0.2130(x - 5.9)^2 + 0.1207(x - 5.9)^3$
Optical/ NIR	$y := (x - 1.82)$
	$a(x) = 1 + 0.17699y - 0.50447y^2 - 0.02427y^3 + 0.72085y^4 + 0.01979y^5 - 0.77530y^6 + 0.32999y^7$
	$b(x) = 1.41338y + 2.28305y^2 + 1.07233y^3 - 5.38434y^4 - 0.62251y^5 + 5.30260y^6 - 2.09002y^7$
IR	$a(x) = 0.574x^{1.61}$
	$b(x) = -0.527x^{1.61}$

A standard approach used in the literature to deredden the data is to estimate the amount of reddening from extinction tables (e.g., Schlegel et al. 1998; Gonzalez et al. 2013; Green et al. 2018) given an object's equatorial coordinates. The tables provide color excess along the line of sight, which, given R_V and distance, can be used to calculate passband extinction $A(\lambda)$. This extinction is then subtracted from photometric observations. These tables are either 2D (integrated along the line of sight) or 3D (provide color excess as a function of distance). The "ultimate" extinction table is expected to come from ESA's *Gaia* mission (Schultheis et al. 2015).

There is a significant problem with this standard approach when applied to eclipsing binary stars: as there are two stars that contribute to total flux and their relative contributions change with phase (i.e., ellipsoidal variations, eclipses, spots, etc.), the amount of reddening also changes, and subtracting a single constant introduces systematic trends as a function of phase. The magnitude of these trends increases with the color difference between the components and with the amount of extinction (Prša & Zwitter 2005).

A much more suitable approach to account for interstellar extinction is to multiply the underlying model atmospheres by monochromatic extinction $A(\lambda)$ and integrate over wavelength:

$$\langle I_{\mathrm{pb,\,norm}}\rangle = \frac{\int_\lambda \mathcal{P}(\lambda)A(\lambda)I_{\lambda,\,\mathrm{norm}}(\lambda)\,d\lambda}{\int_\lambda \mathcal{P}(\lambda)A(\lambda)\,d\lambda}, \tag{5.23}$$

so Equation (4.27) becomes:

$$F_{\mathrm{pb}} = \frac{\mu'}{D^2}\left(\int_\lambda \mathcal{P}(\lambda)A(\lambda)d\lambda\right)\int_{\partial V}\langle I_{\mathrm{pb,\,norm}}\rangle\mu\mathcal{V}(\boldsymbol{r})\mathcal{L}(\mu)dA. \tag{5.24}$$

Thus, we can expand the atmosphere lookup tables by adding another dimension, A_{pb}, and interpolate as before to obtain the correct amount of interstellar extinction added to the model (Jones et al. 2018, in preparation).

To demonstrate the effect of reddening on eclipsing binaries, consider a pair of main-sequence stars with temperatures 6500 and 5500 K, orbiting in a circular orbit with an $i = 90°$ inclination. Say we observe such a binary in Johnson B and V passbands. We will correct for extinction in two ways: by using the standard approach and by using the wavelength-dependent approach. We will neglect all other effects (limb darkening, gravity brightening, reflection effect) so that we evaluate solely the impact of reddening.

The logic is as follows: for the given phase, we compute the effective spectrum of the binary by summing the Doppler-shifted spectra of the visible surfaces of both components. To this intrinsic spectrum we apply interstellar extinction as a function of wavelength. We then multiply this reddened spectrum by the instrumental response function (filter transmission × detector response) and integrate over the passband wavelength range to obtain the flux.

For the standard approach we use the same intrinsic spectrum without applying the wavelength-dependent extinction; instead, we divide the intrinsic spectrum by the extinction constant and integrate as before. We can then readily compare the two fluxes as a function of phase. There is a catch, though: as the Cardelli et al. (1989) formula depends on the wavelength, the extinction constant will also depend on the wavelength. So which wavelength do we choose? We could naively take the effective wavelength of the passband, i.e., the wavelength that splits the passband integral into halves. That is not what we need, though; we need a wavelength that splits the product $\mathcal{P}(\lambda)A(\lambda)I_\lambda(T_{\mathrm{eff}}, \log g, \ldots)$ into halves. To compute this, we need to repeat the same wavelength integration process as before just to get a suitable effective wavelength. Table 5.1 summarizes effective wavelengths computed as a geometric mean of different quantities: passband, reddened passband, intrinsic spectrum, reddened spectrum, and the "true" composite reddened spectrum. Clearly, effective wavelength depends strongly on the effective temperature of the observed object and on the color excess $E(B-V)$ (Figure 5.10). Figure 5.11 depicts the discrepancy between the properly calculated light curve and the one obtained by subtracting a λ_{eff}-calculated constant.

Table 5.1. Summary of Different Approaches to Calculate the Wavelength to Be Used for the Dereddening Constant.

Approach	λ_{eff}	Δm_B	$\epsilon_{\Delta m_B}$
Rigorously calculated value	4470.2 Å	4.03	0.00
Passband transmission	4410.8 Å	4.09	0.06
Passband transmission + reddening law	4452.1 Å	4.05	0.02
Intrinsic spectrum	4436.3 Å	4.06	0.03
Effective (reddened) spectrum	4583.6 Å	3.90	−0.13

λ_{eff} is determined by requiring that the area under the spectrum on both sides is equal. Δm_B is the value of extinction in the B passband, and $\epsilon_{\Delta m_B}$ is the deviation from the rigorously calculated value. All values are calculated for $E(B-V)=1$ at quarter phase.

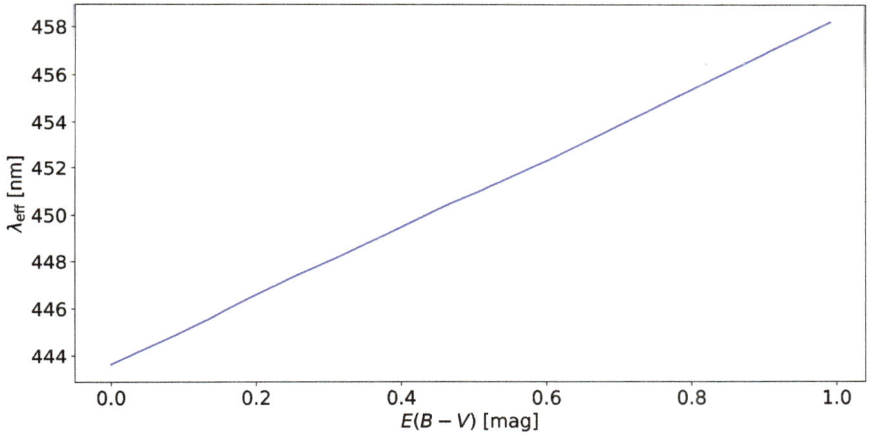

Figure 5.10. Change of the effective wavelength of the Johnson B filter due to reddening of the simulated binary.

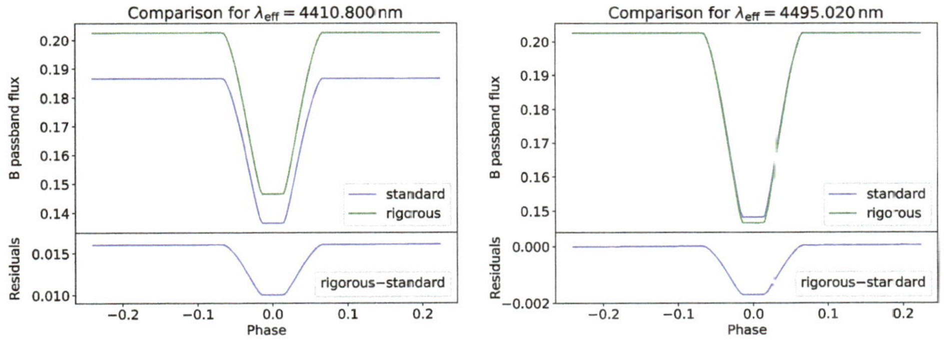

Figure 5.11. Left: discrepancy due to a simplified constant subtraction approach (blue line) compared to the rigorously applied reddening (green line) for a model binary. The subtracted constant was obtained from the effective wavelength ($\lambda_{\text{eff}} = 4410.8$ Å) of the Johnson B transmission curve. Right: overplotted light curves with the subtraction constant calculated for the quarter-phase composite reddened spectrum. There is still a measurable difference in eclipse depth of both light curves. $E(B-V)=1$ is assumed.

Even if the subtraction constant is properly calculated, the light curves will still exhibit measurable differences in both minima (Figure 5.11). This is because of the color index (effective temperature) dependence on phase, most notably during eclipses. For the case at hand, the difference in B magnitude is ~1.75 mmag. For the case of a B0V–K4III binary, the same difference inflates to ~0.2 mag (Prša & Zwitter 2005).

If light curves in three or more photometric passbands are available, it is possible to uniquely determine the color excess value $E(B-V)$ by comparing color indices in and out of eclipse. The reddening may thus be properly introduced as a parameter that can be adjusted during the fitting process.

Atmospheric extinction is an equivalent but a better-posed problem. As interstellar extinction depends on $E(B-V)$, atmospheric extinction depends on air mass, which (unlike $E(B-V)$) is a directly measurable quantity. The downside of atmospheric extinction compared to interstellar extinction is its short-scale time variability: due to the rapidly changing nature of atmospheric convection cells, the presence of cirrus clouds, and water vapor, air mass changes unpredictably during the course of observations.

To model atmospheric extinction, we use the equation triplet for Rayleigh-ozone-aerosol extinction sources given by Forbes et al. (1996) and summarized by Pakštiene & Solheim (2003):

$$k_{\text{total}} = k_{\text{Rayleigh}} + k_{\text{ozone}} + k_{\text{aerosol}}. \tag{5.25}$$

Rayleigh scattering is given by the following expression:

$$k_{\text{Rayleigh}} = k_{\text{Rayleigh}}^{\text{STP}} \left(\frac{P}{P_{\text{STP}}} \right) \left(\frac{T_{\text{STP}}}{T} \right), \tag{5.26}$$

where P and T are the pressure and temperature during observations; $P_{\text{STP}} =$ 101,300 Pa and $T_{\text{STP}} = 273.15$ K are the standard temperature and pressure, respectively. Rayleigh scattering for these standard values is given by:

$$k_{\text{Rayleigh}}^{\text{STP}} = 9.4977 \cdot 10^{-3} e^{-h/7.996} \frac{1}{\lambda^4} \left(\frac{n(\lambda) - 1}{n(\lambda = 1\mu\text{m}) - 1} \right)^2, \tag{5.27}$$

where λ is the wavelength in μm. The formula is normalized to $\lambda = 1\mu$m, with n being the refractive index and the coefficient 9.4977×10^{-3} corresponding to the optical thickness of the atmosphere at $\lambda = 1$ μm. This is adjusted for the height h (in km) of the observatory above sea level, with the density scale height for the lower troposphere assumed to be 7.996 km (Forbes et al. 1996). The normalized index of refraction term may be expressed in terms of wavelength by an empirical relation proposed by Penndorf (1957):

$$\left(\frac{n(\lambda) - 1}{n(\lambda = 1\mu\text{m}) - 1} \right) = 0.23465 + \frac{107.6}{146 - \lambda^{-2}} + \frac{0.93161}{41 - \lambda^{-2}}. \tag{5.28}$$

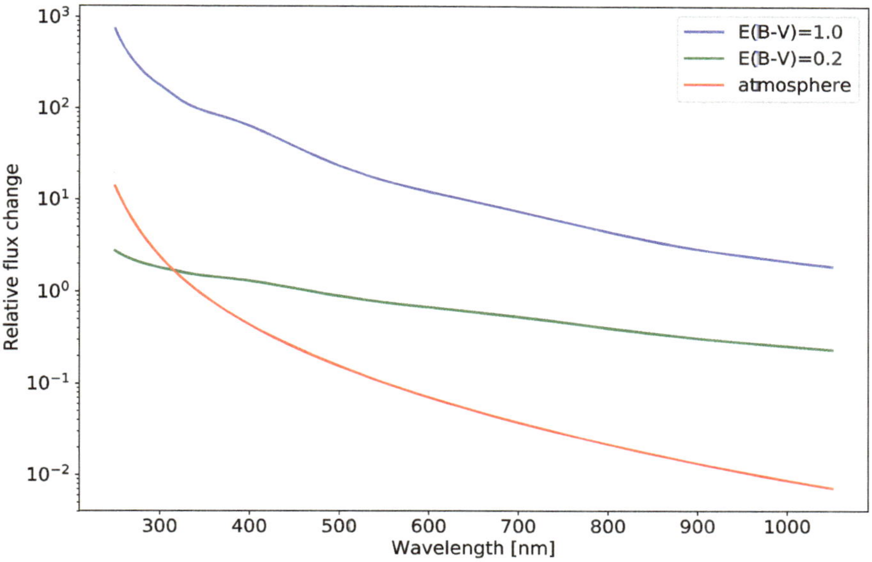

Figure 5.12. Relative flux difference between the intrinsic spectrum and reddened spectrum by $E(B-V) = 1.0$ (blue line), by $E(B-V) = 0.2$ (red line), and by atmospheric extinction (green line). The observatory altitude $h = 0$ km and the zenith air mass are assumed throughout the study. If the color excess is large, reddening dominates the spectrum, but if the color excess is moderate, the blue part of the spectrum is dominated by the Rayleigh scattering.

Ozone extinction depends on the amount α_{ozone} of ozone in the vertical column of the atmosphere with cross section 1 cm^2 and the ozone absorption coefficient $\kappa_{\text{ozone}}(\lambda)$:

$$k_{\text{ozone}} = 1.1\ \alpha_{\text{ozone}}\ \kappa_{\text{ozone}}(\lambda), \tag{5.29}$$

where 1.1 corresponds to the ozone optical thickness at 320 nm. Values of κ_{ozone} are tabulated, e.g., in Cox (2000).

Aerosol scattering on particles with size of the same order of magnitude as the wavelength is the most complicated contribution to the overall atmospheric extinction. It is modeled by the power law:

$$k_{\text{aerosol}} = A\lambda^{-b}, \tag{5.30}$$

where A is the extinction at zenith for $\lambda = 1\ \mu$m and b is the coefficient that depends on the size and distribution of aerosol particles. These two parameters are very difficult to assess; fortunately, the aerosol contribution is usually very small, even negligible (Pakštiene & Solheim 2003). We neglect its influence in our simulation.

In terms of binary stars, atmospheric extinction depends on the wavelength of the observed light and on the air mass of observations, which is of course the same for both binary components; we are thus left only with wavelength dependence at some inferred air mass. We can therefore compare the flux change imposed on the intrinsic spectrum by reddening and by atmospheric extinction (Figure 5.12). For weakly and moderately reddened eclipsing binaries atmospheric extinction dominates the blue parts of the spectrum owing to Rayleigh scattering, but for larger color excesses ($E(B-V) \gtrsim 0.5$) interstellar extinction is the dominant factor throughout the spectrum. Since air mass can be measured reliably from observations of comparison stars, the effect of atmospheric extinction can in principle be removed from the data.

References

Alencar, S. H. P., & Vaz, L. P. R. 1997, A&A, 326, 257

Alencar, S. H. P., Vaz, L. P. R., & Nordlund, Å. 1999, A&A, 346, 556

Bloemen, S., Marsh, T. R., Degroote, P., et al. 2012, MNRAS, 422, 2600

Bloemen, S., Marsh, T. R., Østensen, R. H., et al. 2011, MNRAS, 410, 1787

Boulliau, I. 1667, Ismaelis Bullialdi Ad astronomos monita duo: Primum, de stella noua, quae in collo Ceti ante annos aliquot visa est, Alterum, de neb- ulosa in Andromeda cinguli parte Borea ante biennium iterum orta (Paris: Apud Sebastianum Mabre-Cramoisy)

Cardelli, J. A., Clayton, G. C., & Mathis, J. S. 1989, ApJ, 345, 245

Claret, A. 1999, in ASP Conf. Ser. 173, Stellar Structure: Theory and Test of Connective Energy Transport, ed. A. Gimenez, E. F. Guinan, & B. Montesino (San Francisco, CA: ASP), 277

Claret, A. 2000, A&A, 363, 1081

Claret, A., & Bloemen, S. 2011, A&A, 529, A75

Cox, A. N. 2000, Allen's Astrophysical Quantities, ed. A. N. Cox (4th ed.; New York: AIP Press)

Draine, B. T. 2003, ARA&A, 41, 241

Espinosa Lara, F., & Rieutord, M. 2011, A&A, 533, A43

Espinosa Lara, F., & Rieutord, M. 2012, A&A, 547, A32

Fabricius, J. 1611, De maculis in sole observatis narration (Wittenberga: ETH-Bibliothek Zürich)

Forbes, M. C., Dodd, R. J., & Sullivan, D. J. 1996, BaltA, 5, 281

Gonzalez, O. A., Rejkuba, M., Zoccali, M., et al. 2013, A&A, 552, A110

Green, G. M., Schlafly, E. F., Finkbeiner, D., et al. 2018, arXiv:1801.03555

Gustafsson, B. 1983, MitAG, 60, 139

Hilditch, R. W., Harries, T. J., & Hill, G. 1996, MNRAS, 279, 1380

Horvat, M., Conroy, K. E., Pablo, H., et al. 2018, ApJS, 237, 26

Kippenhahn, R. 1977, A&A, 58, 267

Kron, G. E. 1947, PASP, 59, 261

Loeb, A., & Gaudi, B. S. 2003, ApJL, 588, L117

Lucy, L. B. 1967, ZA, 65, 89

Martynov, D. 1973, SvPhU, 15, 786

Olson, E. C., & Etzel, P. B. 1994, AJ, 108, 262

Pakštiene, E., & Solheim, J.-E. 2003, BaltA, 12, 221

Penndorf, R. 1957, JOSA, 47, 176

Pigulski, A., & Michalska, G. 2002, IBVS, 5218, 1

Prša, A., Conroy, K. E., Horvat, M., et al. 2016, ApJS, 227, 29

Prša, A., & Zwitter, T. 2005, in ESA SP-576: The Three-Dimensional Universe with Gaia, ed. C. Turon, et al. (Paris: Observatoire de Paris-Meudon), 611

Rafert, J. B., & Twigg, L. W. 1980, MNRAS, 193, 79

Roettenbacher, R. M., Kane, S. R., Monnier, J. D., & Harmon, R. O. 2016, ApJ, 832, 207

Rybicki, B. G., & Lightman, P. A. 1979, Radiative Processes in Astrophysics (New York: Wiley)

Schlegel, D. J., Finkbeiner, D. P., & Davis, M. 1998, ApJ, 500, 525

Schultheis, M., Kordopatis, G., Recio-Blanco, A., et al. 2015, A&A, 577, A77
Spruit, H. C., & Weiss, A. 1986, A&A, 166, 167
Strassmeier, K. G. 2009, A&ARv, 17, 251
Trumpler, R. J. 1930, PASP, 42, 267
van Kerkwijk, M. H., Rappaport, S. A., Breton, R. P., et al. 2010, ApJ, 715, 51
Vaz, L. P. R. 1985, Ap&SS, 113, 349
von Zeipel, H. 1924, MNRAS, 84, 665
Warner, B. 2003, Cataclysmic Variable Stars (Cambridge: Cambridge Univ. Press)
Welsh, W. F., Orosz, J. A., Aerts, C., et al. 2011, ApJS, 197, 4
Wilson, R. E. 1990, ApJ, 356, 613
Zucker, S., Mazeh, T., & Alexander, T. 2007, ApJ, 670, 1326

Chapter 6

Advanced Topics

Using the *Kepler* satellite (Borucki et al. 2010), around 3000 eclipsing binary stars have been discovered in its ~105 deg^2 field of view (Prša et al. 2011; Kirk et al. 2016; Abdul-Masih et al. 2016). These systems were observed for 4 yr with an unprecedented photometric precision (~30 ppm) and a ~92% duty cycle. The study of binary star physics has undergone a monumental leap thanks to *Kepler*. For example, (1) Doppler boosting has been detected for the first time (Bloemen et al. 2012); (2) spin–orbit misalignment (the anomalous Rossiter effect) has been observed in numerous hot stars and a few cool stars (Mazeh et al. 2015); (3) a dozen circumbinary planets have been discovered (Doyle et al. 2011; Orosz et al. 2012b; Welsh et al. 2012, and others that followed), including the multiple circumbinary planet system *Kepler*-47 (Orosz et al. 2012a); (4) thermally bloated low-mass white dwarfs in eclipsing binary systems were discovered that put our understanding of white dwarf evolution to the test (Rappaport et al. 2015); and (5) asteroseismic signals, including pressure (p), gravity (g), and hybrid modes, were found in eclipsing binary light curves (Huber 2015), enabling us to use asteroseismic and binary modeling tools in conjunction. The *Kepler* data have proven to be a gold mine for stellar astrophysics.

Gaia (Prusti et al. 2016), the European Space Agency's cornerstone mission, was launched in December 2013, and it started its science operation in mid-2014. It carries an astrometric instrument that is composed of two telescopes, 14 CCDs devoted to sky mappers, and 62 CCDs in the focal plane where the two fields of view are combined to enable astrometric measurements. *Gaia* is also equipped with a photometric instrument; it uses the same optical design, with an addition of two fused-silica prisms dispersing light into low-resolution spectrophotometric beams. The dispersers work in different wavelength regions: BP (blue photometer) operates in the 330–680 nm range, while RP (red photometer) operates in the 640–1050 nm range. Finally, *Gaia* also carries a spectroscopic instrument for the measurement of radial velocities. Its resolving power is ~11,700, and its wavelength range spans 845–872 nm, covering the Ca II triplet. *Gaia* will deliver astrometric measurements at a microarcsecond level, photometric measurements at a millimagnitude level, and spectroscopic (radial velocity) measurements at a 0.5 kms^{-1} level for the bright end of the ~1 billion stars observed. It is safe to say that *Gaia* is revolutionizing all-sky surveys.

The High Accuracy Radial velocity Planet Searcher (HARPS; Mayor et al. 2003) is an unparalleled spectrometer that can attain a radial velocity accuracy of 1 $\mathrm{m\ s}^{-1}$. It is a fiber-fed, cross-dispersed echelle spectrograph that operates in the 72nd order, covering the spectral region between 380 and 690 nm and yielding resolving power $R \sim 115{,}000$. One of the major achievements with this instrument is a volume-limited census of 582 FGK stars (Sousa et al. 2011), delivering the largest sample of spectroscopic stellar parameters of the solar neighborhood.

The basic considerations discussed in this book so far make the description of real stellar systems complex and convoluted. As established above, modern data reach unprecedented precision on the level of μas in astrometry, ppm in photometry, and $\mathrm{m\ s}^{-1}$ in spectroscopy, and the level of detail that emerges from these observations allows us to probe second-order effects routinely. In this chapter we develop a formalism for some of these effects and discuss future prospects for effects where our computational ability is not quite up to the task yet.

6.1 Misaligned Stellar Rotation Axes

It seems natural to expect a high degree of alignment between the axes of rotation (spin angular momenta) and axes of revolution (orbital angular momenta) in stellar and planetary systems. The angle between the axes of rotation and revolution (or, conversely, between the equatorial and orbital planes) is called *axial tilt* or *obliquity*. Given our understanding of the nebular theory that describes the contraction of the protostellar cloud into stars and planetary systems, we typically invoke the conservation of angular momentum of the cloud to predict such alignment; any deviation is attributed to close encounters and collisions at an early stage of formation. Taking a quick look at our solar system, it becomes abundantly clear that this premise is not warranted. Starting with the Sun, its equator is tilted to the ecliptic by 7.25°. Earth's equator is tilted on average 23.44°, and it precesses at a rate of $\sim 50'' \mathrm{yr}^{-1}$, so obliquity changes as a function of time. Solar system planets all orbit within ~7° of the ecliptic plane, but their rotational axes are nowhere near aligned (see Table 6.1; Astronomical Almanac 2017). Transiting exoplanet host stars exhibit a wide range in their obliquities, from completely aligned (i.e., HD 189733; Winn et al. 2006), to moderately aligned (i.e., XO-3 at 37°; Hirano et al. 2011), to perpendicular (WASP-7; Albrecht et al. 2012) and even retrograde (WASP-17; Anderson et al. 2010). Giant extrasolar planets on very close, eccentric orbits (the so-

Table 6.1. Axial Tilts (Obliquities) for Solar System Planets

Planet	Mercury	Venus	Earth	Mars	Jupiter	Saturn	Uranus	Neptune
Obliquity	0.04°	2.64°	23.44°	25.19°	3.12°	26.73°	82.23°	28.34°

called hot and warm Jupiters) also demonstrate a large range and oscillatory behavior in their obliquities (Dawson & Chiang 2014). Two good reviews on the methodology and results of obliquity measurements in exoplanet systems are done by Winn & Fabrycky (2015) and Triaud (2017). Lastly, there are two shining examples for misalignment among eclipsing binary stars: DI Her (Philippov & Rafikov 2013) and CV Vel (Albrecht et al. 2014). DI Her is the current record holder with the sky-projected spin–orbit angles of $\beta_p = 72° \pm 4°$ and $\beta_s = -84° \pm 8°$ for the primary and secondary star, respectively, while CV Vel features a misaligned primary at $\beta_p = -52° \pm 6°$ and an aligned secondary at $\beta_s = 3° \pm 7°$.

The most commonly used method to measure misalignment is to acquire spectroscopic measurements of a binary star during the eclipse. Spectral line profiles and measured radial velocities are determined by the brightness distribution across the visible surfaces of the stars, and this distribution depends substantially on the degree of misalignment. The deviation from the aligned case in radial velocity curves is called the *anomalous* Rossiter–McLaughlin effect.

In Chapter 3 we derived a general expression for the surface potential that includes obliquity, Equation (3.29). Figure 6.1 depicts equipotential surfaces for a binary star with $q = 0.25$ that correspond to the aligned case (left) and to the case where the primary is misaligned by 0.3π (right). The interactive version of the book cycles through the $[0, 2\pi]$ range of obliquities.

Misalignment has been studied in detail by Limber (1963), Kruszewski (1967), and Avni & Schiller (1982); the theoretical framework we discuss here follows closely the work of Horvat et al. (2018).

6.1.1 Generalized Lagrangian Points

The first generalization of the potential, Equation (3.31), introduces asynchronous rotation while retaining spin–orbit alignment. There are still five solutions to Equation (3.45), and their analysis can be done along the same lines as we did for

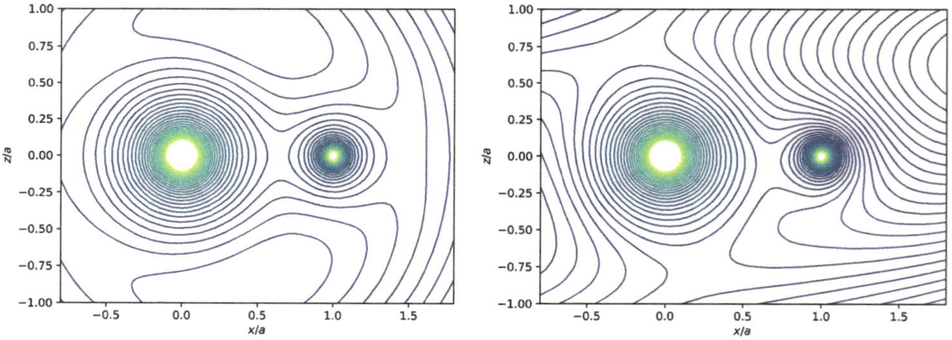

Figure 6.1. Equipotential surfaces of the potential for the binary system with $q = 0.25$. The surfaces in the left panel correspond to a fully aligned case, and the surfaces in the right panel correspond to a misaligned primary star with $\beta_1 = 0.3\pi$ and an aligned secondary star. The interactive version of the book cycles through β in the $[0, 2\pi]$ range. The animation is available online at https://doi.org/10.1088/978-0-7503-1287-5.

a synchronous orbit in Section 3.2.2. These generalized points, denoted $G_i(q, F, \delta)$, $i = 1 \ldots 5$, are identical to Lagrangian points L_i, $i = 1 \ldots 5$, when $F \to 1$. This generalization has been presented in detail by Kallrath & Milone (2009), and we will not repeat it here.

Of utmost interest is the study of generalized points in misaligned systems. The principal legwork was done by Avni & Schiller (1982) and by Horvat et al. (2018). Consider the generalized potential given by Equation (3.29); that form can be transformed to the generalized dimensionless form:

$$\begin{aligned}\Omega(\boldsymbol{\rho};\, q,\, F,\, \delta,\, \hat{\boldsymbol{S}}) &= \frac{1}{\rho} + q\left(\frac{1}{\|\boldsymbol{\rho} - \delta\hat{\mathbf{\imath}}\|} - \frac{\hat{\mathbf{\imath}}\cdot\boldsymbol{\rho}}{\delta^2}\right)\\ &+ \frac{1}{2}(1 + q)F^2\|(\hat{\boldsymbol{S}}\cdot\boldsymbol{\rho})\hat{\boldsymbol{S}} - \boldsymbol{\rho}\|^2,\end{aligned} \tag{6.1}$$

where $\hat{\boldsymbol{S}}$ is the unit spin vector, $q \equiv M_2/M_1$ is the mass ratio, $F = \omega_S/\omega_L$ is the synchronicity parameter, and $\delta \equiv d/a$ is the fractional instantaneous separation. We can characterize the misalignment of $\hat{\boldsymbol{S}}$ by rotating the canonical coordinate system into the corotating coordinate system (see Figure 3.3) so that $\hat{\boldsymbol{S}}$ lies in the $x'z'$ plane:

$$\hat{\mathbf{\imath}}' = \hat{\mathbf{\imath}}, \quad \hat{\mathbf{\jmath}}' = \hat{\mathbf{\jmath}}\cos\alpha + \hat{\mathbf{k}}\sin\alpha, \quad \hat{\mathbf{k}}' = -\hat{\mathbf{\jmath}}\sin\alpha + \hat{\mathbf{k}}\cos\alpha, \tag{6.2}$$

where $\alpha = -\arctan(\hat{\boldsymbol{S}}\cdot\hat{\mathbf{\jmath}}/\hat{\boldsymbol{S}}\cdot\hat{\mathbf{k}})$. The spin vector in the new basis is written as:

$$\hat{\boldsymbol{S}} = \hat{\mathbf{\imath}}'\sin\beta + \hat{\mathbf{k}}'\cos\beta, \quad \beta = \arcsin(\hat{\boldsymbol{S}}\cdot\hat{\mathbf{\imath}}'). \tag{6.3}$$

It proves useful to do one last rescaling of the potential; if $\rho' \to \rho'\delta$, $\Omega \to \Omega\delta$ and we introduce $b \equiv (1 + q)F^2\delta^3$, then the potential in the corotating coordinate system can be written as:

$$\begin{aligned}\Omega(\boldsymbol{\rho}';\, q,\, b,\, \beta) &= \frac{1}{\rho'} + q\left(\frac{1}{\|\boldsymbol{\rho}' - \hat{\mathbf{\imath}}'\|} - \hat{\mathbf{\imath}}'\cdot\boldsymbol{\rho}'\right)\\ &+ \frac{1}{2}b[(x'\cos\beta - z'\sin\beta)^2 + y'^2].\end{aligned} \tag{6.4}$$

To obtain generalized Lagrangian points $\boldsymbol{G}_i$, we need to solve $\nabla\Omega(\boldsymbol{G}_i) = 0$. To this end we follow the same procedure as we did in Section 3.2.2; the result is a bit trickier to interpret now because the number of Lagrange points is no longer necessarily 5. The above condition for equilibrium points can be written as:

$$\begin{aligned}&\left(b\begin{bmatrix}\cos^2\beta & 0 & -\sin\beta\cos\beta\\ 0 & 1 & 0\\ -\sin\beta\cos\beta & 0 & \sin^2\beta\end{bmatrix} - \left(\frac{1}{\rho_1^3} + \frac{q}{\rho_2^3}\right)\boldsymbol{I}\right)\boldsymbol{G}_i\\ &+ q\left(\frac{1}{r_2^3} - 1\right)\hat{\mathbf{\imath}} = 0,\end{aligned} \tag{6.5}$$

where $\boldsymbol{I}$ is the identity matrix, $\rho_1 \equiv \|\boldsymbol{G}_i\|$, and $\rho_2 \equiv \|\hat{\mathbf{i}} - \boldsymbol{G}_i\|$. As before, we make a distinction regarding the location of the equilibrium points. They may lie in the $x'z'$ plane (generalized L_1, L_2, and L_3) or outside the $x'z'$ plane (generalized L_4 and L_5). After some lengthy algebraic manipulation, we arrive at the following generalizations:

- Generalized points $\boldsymbol{G}_4$ and $\boldsymbol{G}_5$ are real for $b - q \geqslant (\sqrt{2}\,|\cos\beta|)^{-3}$ and imaginary otherwise; their coordinates are:

$$\left(\frac{1}{2}(b-q)^{-2/3},\ \pm\sqrt{(b-q)^{-2/3} - \frac{\frac{1}{4}(b-q)^{-4/3}}{\cos^2\beta}},\ -\frac{1}{2}(b-q)^{-2/3}\tan\beta\right). \tag{6.6}$$

- Generalized points in the $x'z'$ plane are difficult to compute. The main reason is the nonlinearity of the parameter space. Figure 6.2 demonstrates the richness of solutions for equilibrium points as a function of q, F, and β. In some circumstances we have a single equilibrium point, while in others we have all three. Horvat et al. (2018) developed a dedicated numerical algorithm to find the generalized equilibrium points in the $x'z'$ plane that is based on evaluating $d\boldsymbol{G}_i(\beta)/d\beta$,

$$\frac{d\boldsymbol{G}_i(\beta)}{d\beta} = -[(\nabla \otimes \nabla)\Omega(\boldsymbol{G}_i(\beta); q, b, \beta)]^{-1} \cdot \nabla\frac{\partial\Omega(\boldsymbol{G}_i(\beta); q, b, \beta)}{\partial\beta}, \tag{6.7}$$

and integrating it numerically on the $[0, \beta]$ range with the initial condition $\boldsymbol{G}_i(\beta = 0) = \boldsymbol{G}_i^{\text{aligned}}(q, b)$. This can be done for as long as the Hessian, $(\nabla \otimes \nabla)\Omega$, is nonsingular. The bottom line is that the existence and behavior of equilibrium points in the $x'z'$ plane strongly depend on the topological parameters and need to be evaluated for each circumstance separately.

6.1.2 Orbital Misalignment in the Plane of Sky

General transformations between the canonical coordinate system and the plane of sky have been derived in Section 3.3.1, and we can readily apply them here. The direction of orbital angular momentum is given by:

$$\hat{\boldsymbol{L}} = \boldsymbol{R}_{uw}(i, \Omega)\hat{\boldsymbol{w}} = [-\sin i \sin\Omega, \sin i \cos\Omega, \cos i]^T, \tag{6.8}$$

where Ω is the longitude of the ascending node and i is the inclination. Similarly, the spin unit vector can be transformed as:

$$\hat{\boldsymbol{S}} = \boldsymbol{R}_{uw}(i', \Omega')\hat{\boldsymbol{w}} = [-\sin i' \sin\Omega', \sin i' \cos\Omega', \cos i']^T, \tag{6.9}$$

where i' and Ω' are given with respect to the corotating coordinate system. Instead of thinking of the angles with respect to misaligned coordinate systems, it is much simpler to define their differences as new parameters:

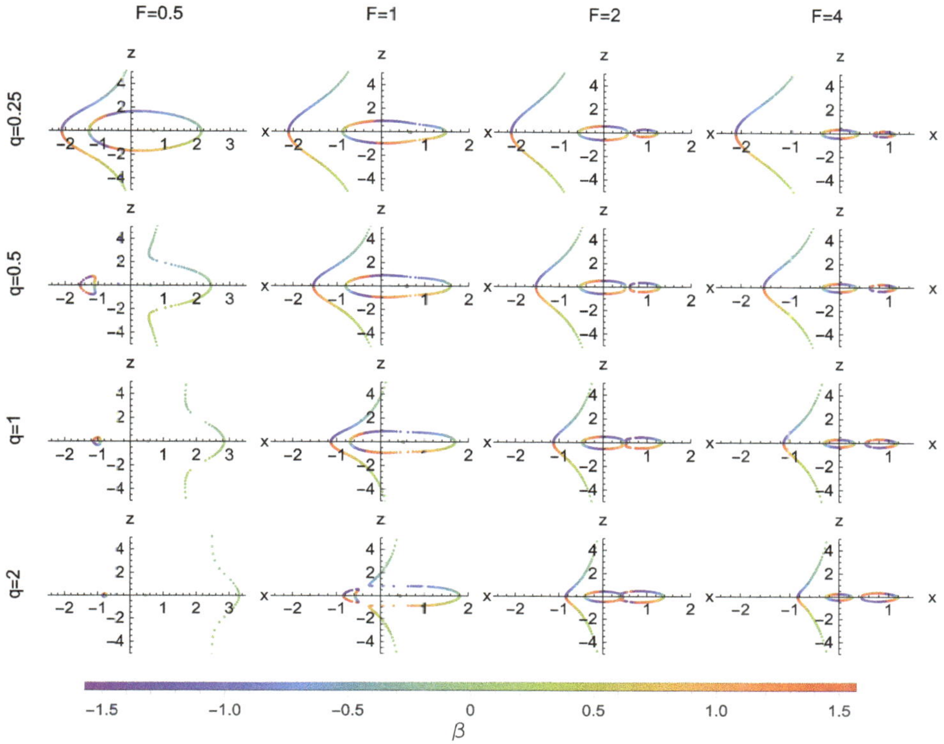

Figure 6.2. Generalized Lagrangian points as a function of q (vertical), F (horizontal), and β (color-coded). The existence of equilibrium points depends strongly on all parameters.

$$\text{pitch:} \quad \xi \equiv \Delta i = i' - i; \qquad \text{yaw:} \quad \eta = \Delta\Omega = \Omega' - \Omega. \tag{6.10}$$

The spin vector can then be described by pitch and yaw uniquely as:

$$\hat{S} = \boldsymbol{R}_w(-\varpi)\boldsymbol{R}_u(i)\boldsymbol{R}_w(\eta)\boldsymbol{R}_u(-i - \xi)\hat{\boldsymbol{w}}, \tag{6.11}$$

where $\varpi \equiv \omega + \Delta + \upsilon$, ω being the argument of periastron, Δ being the positional offset (π for the primary star and 0 for the secondary star), and υ being the true anomaly.

6.1.3 Misalignment in Practice

To demonstrate the effect of misalignment on light curves, radial velocity curves, and line profiles, consider a system described by parameters given in Table 6.2 (adapted from Horvat et al. (2018)). Figure 6.3 depicts the comparison between aligned and misaligned observables; the effect of misalignment is clearly pronounced in both light curves and radial velocity curves. A telltale sign of misalignment is the asymmetry in light curves, although asymmetries can arise from other physical effects as well, such as ellipsoidal variation and reflection in eccentric systems. Radial velocity effects are pronounced in the eclipses (the Rossiter–McLaughlin

Table 6.2. Principal Parameters of the Toy Model for a Misaligned Binary Star

Parameter	Primary Star	System	Secondary Star
Semimajor axis a ($R_\odot$)		3.98	
Period P (days)		0.65	
Mass ratio q		0.7	
Eccentricity e		0	
Inclination i (deg)		80	
Long. of ascending node Ω (deg)		0	
Systemic velocity γ (km s^{-1})		0	
Atmosphere	Blackbody		Blackbody
Equivalent radius R ($R_\odot$)	1.3		0.8
Effective temperature $T_{\rm eff}$ (K)	6500		5500
Synchronicity parameter F	1		3.61
LD model	Logarithmic		Logarithmic
LD coefficient x_{LD}	0.5		0.5
LD coefficient y_{LD}	0.5		0.5
Gravity darkening $\beta_{\rm grav}$	0.32		0.32
ξ (deg)	0		45
η (deg)	0		45

effect). Line profiles are depicted for several phases in the aligned system (dashed line), the misaligned system (solid line), and the purely spherical system (dot-dashed line). As expected, the differences are the largest during eclipses, as the main driver for the line profile is the sum of the local intensities weighted by the projected surface element area. Even outside the eclipses, though, the widths of the line profiles can be substantially different (top left panel) because of the modified surface brightness distribution across the disk of the secondary star. The Rossiter–McLaughlin effect is symmetric for the aligned case but is generally asymmetric in the misaligned case. This is depicted in Figure 6.3: the effect is symmetric near time stamp 0, where the aligned primary star is eclipsed, and asymmetric near time stamp 0.32, where the misaligned secondary star is eclipsed.

6.2 Multiple Systems

One of *Kepler*'s prime discoveries regarding binary systems is that ~20% show evidence of tertiary components (Conroy et al. 2014; Orosz 2015). This figure is derived by measuring eclipse timing variations (ETVs)—deviations from the expected timing of eclipses due to gravitational third-body interactions, chromospheric activity, apsidal motion, mass transfer, etc.—for the entire *Kepler* eclipsing binary catalog. Of those ~20%, more than half exhibited obvious periodicities; 16 systems showed significant eclipse depth changes and orbital perturbations on timescales of a single fly-by, and 25 eclipsing binaries featured additional eclipse

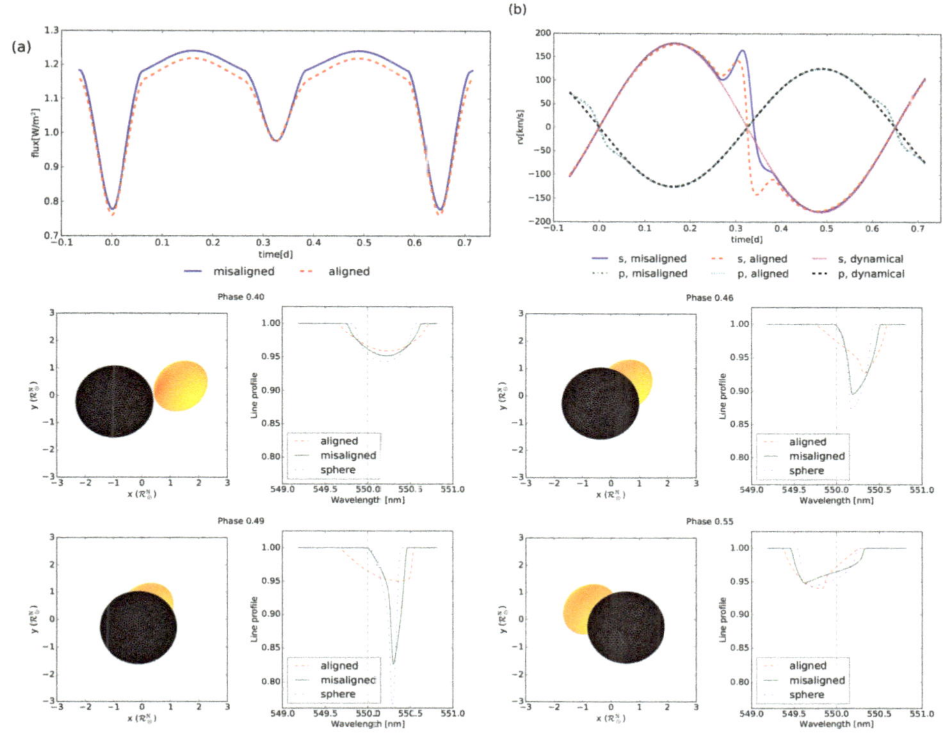

Figure 6.3. Top: comparison of observables for the aligned and misaligned binary system. The parameters used are given in Table 6.2. Labels p and s denote primary and secondary star, respectively. The dynamical (theoretical) radial velocity curves are also depicted for comparison. Bottom: plane-of-sky-projected stellar meshes and the corresponding line profiles for four separate phases. The misaligned star is color-coded by surface brightness. The dashed gray line denotes the rest wavelength of the line; the solid green, dashed red, and dot-dashed blue lines represent the line profiles of the misaligned, aligned, and spherical star systems, respectively.

events. Other ETVs were caused either by the dynamical perturbations of the orbit or by spots affecting the photocenter of the star and causing spurious ETV shifts (Orosz 2015). To properly put these findings in perspective, we need to first discuss the theoretical background of eclipse timing variations.

6.2.1 Eclipse Timing Variations

Because of the finite speed of light, eclipses do not occur at the precise time of the geometric alignment of the two components with the line of sight, t_{geom}, but slightly earlier for the star in the back and slightly later for the star in the front (Irwin 1952). This is called the Rømer delay or, in the field of binary/multiple stars, the light-travel time effect (LTTE).

Interactions in multiple stellar systems make the dynamical properties of eclipse timing much more complex. Suppose we have a triple system with an eclipsing binary orbiting a system's center of mass with a tertiary component. The motion of the binary star's barycenter introduces a change in the distance between the binary components and the observer as it orbits around the common center of mass. At the same time, if the components are sufficiently close, gravitational interactions can change orbital parameters on the scale of a single fly-by. We distinguish three types of contributions to eclipse timing variations:

INTRINSIC: these are caused by the transfer of angular momentum and/or energy between the components or dissipation from the system; this typically happens in semi-detached or contact configurations during conservative mass transfer, ejections and eruptions during nonconservative mass transfer, etc.

DYNAMICAL: when inner and outer orbits are not hierarchical, dynamical interactions between individual components change orbital parameters (inclination, eccentricity, semimajor axis, argument of periastron, etc.).

GEOMETRIC: due to the LTTE as the inner binary's barycenter orbits around the system's center of mass.

ETVs in triple systems have been studied extensively by Borkovits et al. (2003, 2007) and, for the case of exoplanetary transits, Borkovits et al. (2011). In the absence of intrinsic and dynamical contributions, Rappaport et al. (2013) formulated an elegant expression for the LTTE contribution to ETVs:

$$\Delta t_{\mathrm{LTTE}} = A_{\mathrm{LTTE}}[(1 - e_3^2)^{1/2} \sin \upsilon_3 \cos \omega_3 + (\cos \upsilon_3 - e_3) \sin \omega_3], \tag{6.12}$$

where A_{LTTE} is the amplitude of the variation, and e_3, υ_3, and ω_3 are orbital eccentricity, true anomaly, and argument of periastron of the outer orbit, respectively. The amplitude is given by:

$$A_{\mathrm{LTTE}} = \frac{G^{1/3}}{c(2\pi)^{2/3}} \left[\frac{M_3}{M_{123}^{2/3}} \sin i_3 \right] P_3^{2/3}, \tag{6.13}$$

where M_3 and P_3 are the mass and the orbital period of the third body, and M_{123} is the total mass of the triple system. This form of the equation is convenient because it can be fit directly to the ETV data.

Another source of timing variations that was only recently studied in detail are the barycentric and asymmetric transverse velocities (BATVs; Conroy et al. 2018). This is another nonrelativistic effect that arises from the finite speed of light. If the barycenter of the binary system moves with respect to the observer, and if the velocity has a nonzero transverse (plane-of-sky) component, the measured eclipse times will be affected by that motion. If the motion is constant, such as proper motion on the plane of sky, the BATV effect will be constant and hence difficult or even impossible to measure, but if the velocity vector varies in time, sequential eclipses will exhibit a nonlinear behavior. In practice, the BATV source becomes relevant in eccentric binaries, binaries with apsidal motion, systems with changing inclination, and multiple systems.

Consider a triple system depicted in Figure 6.4. It consists of a massive star (drawn in red) in orbit with a tight binary, which we will call the *inner* binary; its

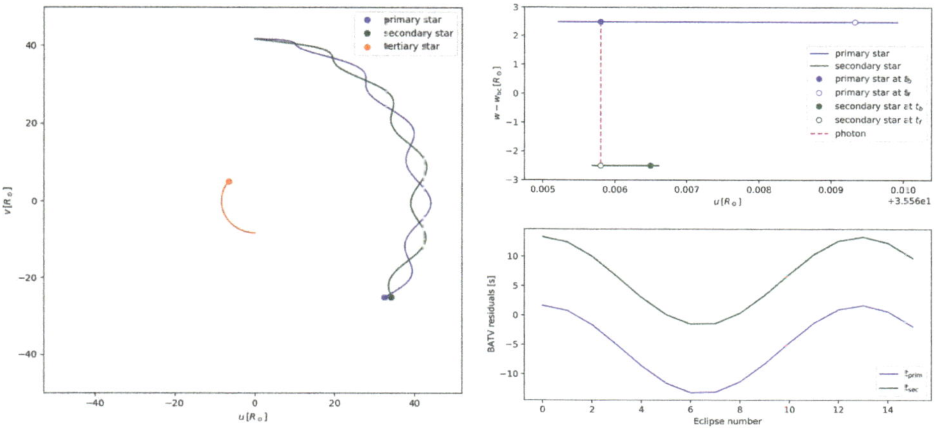

Figure 6.4. Dynamics of the mock triple system. The inner binary has a mass ratio $M_2/M_1 = 1$, a 1-day orbital period, and inclination $i = 90°$. The outer binary has a mass ratio $M_3/(M_1 + M_2) = 5$, a 12.91-day orbital period, and inclination $i = 0°$. Left: uv plane of sky. The orbits of the primary and secondary star interweave on the plane of sky because of the nonzero barycentric motion of the inner binary, causing eclipse timing variations. Top right: depiction of a single primary eclipse event in the uw plane. Filled circles denote the positions of stars at time t_b when the photon was emitted, and open circles denote the positions of stars at time t_f when the photon was intercepted. Bottom right: BATV residuals as a function of the consecutive eclipse number. An animation is available online at https://doi.org/10.1088/978-0-7503-1287-5.

mass ratio is $q_{\text{inner}} = 1.0$, its orbital period is 1 day, and its inclination is 90°. The outer binary's mass ratio is $q_{\text{outer}} = 5.0$, its orbital period is 12.91 days, and its inclination is 0°. This setup was chosen deliberately so that there are no light-travel time contributions other than the BATV. The left panel in the figure depicts the orbits of all three stars in the uv plane of sky. Geometric eclipses would occur at the intersections of the blue and green lines, but because of the net barycentric motion of the inner binary and the finite speed of light, actual eclipses are shifted in time. This is highlighted in the right two panels: the top panel shows the difference in the traversed paths of the primary and the secondary star for one eclipse event, while the bottom panel shows the net BATV effect over 15 eclipse cycles of the inner binary. The peak-to-peak amplitude in this mock situation is ~15 s.

This effect arises from the photon flight length change due to the transverse component of the binary's barycenter velocity. The radial component also plays a role, of course, but that is already accounted for in the traditional LTTE. Given the equations of motion that describe the position of the barycenter as a function of time, $\boldsymbol{r}_{\text{bc}}(t)$, the position of the primary star as a function of time, $\boldsymbol{r}_b(t)$, and the position of the secondary star as a function of time, $\boldsymbol{r}_f(t)$ (indices b and f denote front and back, respectively), we can write the condition for eclipse as a set of two equations:

$$u_b(t_b) = u_f(t_f) \tag{6.14}$$

and

$$w_b(t_b) - w_f(t_f) = c(t_b - t_f). \tag{6.15}$$

Here u_b and u_f are transverse positions of the primary and secondary stars in the plane of sky, w_b and w_f are the positions of the stars along the line of sight, and t_b and t_f are the times of photon emission by the primary star and photon interception by the secondary star, respectively. Equation (6.14) describes the condition that the emitting star and the intercepting star need to be aligned at the time of photon emission and interception, while Equation (6.15) relates the distance of the photon travel and distance between the stars. Solving these two equations in general requires an iterative process that can be outlined as follows:

- pick a time stamp t_b and compute $\boldsymbol{r}_b(t_b)$;
- solve Equation (6.14) for t_f; in most cases this needs to be done iteratively;
- given t_f, compute $\boldsymbol{r}_f(t_f)$;
- given $\boldsymbol{r}_b(t_b)$ and $\boldsymbol{r}_f(t_f)$, calculate the difference between both sides of Equation (6.15), $\Delta = c(t_f - t_b) - w_f(t_f) - w_b(t_b)$;
- iterate the scheme over t_b until $\Delta \to 0$ to a required level of precision.

The results depicted in Figure 6.4 have been computed following this scheme. As we have made no approximations, this scheme, albeit iterative, is *not* approximate; arbitrary accuracy can be attained given the tolerance of the iterative method employed.

In many cases it is appropriate to simplify the above treatment by making the following assumptions:

1. the barycenter of the inner binary does not move in the w-direction during the photon travel time;
2. the components of the inner binary do not move in the w-direction during the photon travel time; and
3. the stars move at a constant speed during the photon travel time.

These assumptions are applicable for systems with orbital periods longer than ~1 day: the systematic error that arises from these assumptions for the showcase depicted in Figure 6.4 is ~1% of the BATV amplitude.

Under these assumptions, Equations (6.14) and (6.15) can be solved analytically, with no need to iterate. For Keplerian orbits, the amplitude of the BATV can be approximated by (Conroy et al. 2018):

$$\Delta t_{\mathrm{BATV}} = \frac{w_f(t_f) - w_b(t_b)}{c}\left(\frac{q-1}{q+1} - \frac{1}{q+1}\frac{v_{x,\,\mathrm{bary}}}{v_{x,\,f}}\right), \tag{6.16}$$

where $q = M_f/M_b$ is the mass ratio. Conroy et al. (2018) provide approximation equations for several specific cases, including a circular binary system, an eccentric binary system with apsidal motion, a binary system with a changing inclination, and a hierarchical triple system.

The net effect on eclipse timing can then be written as:

$$t_{\mathrm{obs}} = t_{\mathrm{geom}} + \Delta t_{\mathrm{LTTE}} + \Delta t_{\mathrm{BATV}}, \tag{6.17}$$

where $t_{\rm geom}$ may need to be further corrected for any dynamical perturbations of orbital elements, $t_{\rm geom} = t_{\rm ephem} + \Delta t_{\rm dyn}$.

6.3 Granulation

Convection—a cyclic macroscopic motion of the gas—is an exceedingly simple concept with an exceedingly complex manifestation. As a blob of gas inside the star is displaced from its original position, gravity and buoyancy compete to either restore the blob to the original position (gravity wins) or carry the blob away toward the surface (buoyancy wins). Under the hood of this gravity–buoyancy interplay is the temperature gradient between the heat source (inner layer) and heat sink (outer layer). If it exceeds the critical value set by the adiabatic expansion, dynamical instability ensues that causes convection. In more rigorous terms (Kippenhahn & Weigert 1994), convection sets in when the radiative gradient $\nabla_{\rm rad}$ (a spatial gradient in temperature) exceeds the adiabatic gradient $\nabla_{\rm ad}$ (adiabatic temperature variation in a gas bubble as it undergoes a pressure change):

$$\nabla_{\rm rad} \equiv \frac{3}{16\pi acG} \frac{\kappa lp}{mT^4} > \nabla_{\rm ad}, \tag{6.18}$$

where a is radiation density constant, c is the speed of light, G is the universal gravitational constant, κ is opacity, l/m is specific local luminosity, p is pressure, and T is temperature. This is known as the Schwarzschild criterion, and it assumes homogeneous composition, $\nabla_\mu \equiv 0$. Neglecting ∇_μ in general is not quite fair, as nuclear reactions produce heavier elements in the core, resulting in a gradient. If ∇_μ were thrown into the mix, the convection criterion would become $\nabla_{\rm rad} > \nabla_{\rm ad} + \nabla_\mu$, known as the Ledoux criterion. However, convection in outer layers, far from the core, mixes the gas well, and $\nabla_\mu \approx 0$ is a very good assumption (Pols 2011).

Equation (6.18) identifies three regimes in which convection occurs:

1. for large values of κ, i.e., in opaque regions of the star. As opacity increases with decreasing temperature, this accounts for outer layers of Sun-like and cooler stars;
2. for large values of l/m, i.e., in regions with a large energy flux. As $l/m \to \epsilon_{\rm nuc}$ toward the center of the star, this accounts for stellar cores of massive stars;
3. for small values of $\nabla_{\rm ad}$, i.e., in partial ionization zones at relatively low temperatures. This accounts for surface convection even if opacity is not sufficiently large. In consequence, stars of all masses have shallow surface convection zones at temperatures where helium and hydrogen are partially ionized (Pols 2011). For the purpose of observing convection, we are naturally interested in the outermost layers of the stellar atmospheres.

To date, granulation has been directly observed in two objects: our Sun (Herschel 1801) and Betelgeuse, a red supergiant (Chiavassa et al. 2010). While the Sun can be easily resolved, the image of Betelgeuse had to be reconstructed from interferometric observations, c.f. Figure 6.5. Both images depict clear signs of convection cells, yet their characteristic sizes differ by roughly four orders of magnitude: a typical granule

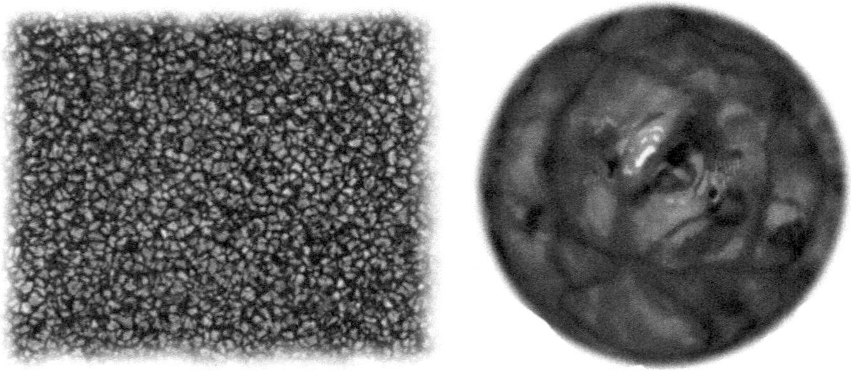

Figure 6.5. Left: near-IR image of solar granulation. Image taken at the Big Bear Solar Observatory in California at 1.56 μm with the 65 cm vacuum reflector. Credit: Big Bear Solar Observatory/New Jersey Institute of Technology. Right: reconstructed image from interferometric observations of Betelgeuse (Chiavassa et al. 2010), a red supergiant. Credit: Chivassa et al., A&A, 515, A12, 2010, reproduced with permission © ESO. These are the only two direct detections of granulation to date.

on the Sun is ~1500km in size and disperses in 8–20 minutes (Bahng & Schwarzschild 1961); a typical granule on Betelgeuse is $\sim 7.0 \times 10^7$ km and is long-lived. Amazingly, though—as we demonstrate below—the characteristics of these cells scale reasonably well across these 4+ orders of magnitude.

That said, *indirect* convection signatures abound. Because of the quasi-periodic nature and quasi-uniform distribution of granules across the surface of the star, convection presents itself as quasi-stochastic noise (Harvey 1985). The nature is quasi-stochastic because granulation causes both temporal and spatial evolution of the surface velocity field, and autocorrelation of the time evolution can be approximated by an exponential decay with a decay constant τ. The background noise spectrum can then be modeled as the sum of Harvey functions:

$$P(\nu) = \frac{4\sigma^2\tau}{1 + (2\pi\nu\tau)^\alpha} \equiv \frac{P_\nu}{1 + (2\pi\nu\tau)^\alpha}, \tag{6.19}$$

where σ is the characteristic amplitude of granulation and α is the characteristic slope of the decay ($\alpha = 2$ for a Lorentzian curve; Mathur et al. 2011). $P_\nu \equiv 4\sigma^2\tau$ is the total power of the granulation signal. Whatever the number of features that contribute to the overall granulation, that is how many Harvey functions we need to include. It is typical to account for granulation, mesogranulation, supergranulation, and active regions, each with a corresponding value of τ, σ, and α; the typical orders of magnitude for the decay constants are τ = 372, 10^4, 10^5, and 10^6 s, respectively (Harvey 1985). It is thus common to neglect long timescales and model the net background noise as a superposition of instrumental noise $P_{\rm instr}$, granulation noise (first term in square brackets), and a power excess hump due to stochastic p-mode pulsations (second term in square brackets):

$$P(\nu) = P_{\rm instr} + \eta(\nu)^2\left[\sum_i \frac{4\sigma_i^2\tau_i}{1 + (2\pi\nu\tau_i)_i^\alpha} + P_g\, e^{-\frac{(\nu-\nu_{\rm max})^2}{2\sigma^2}}\right]. \tag{6.20}$$

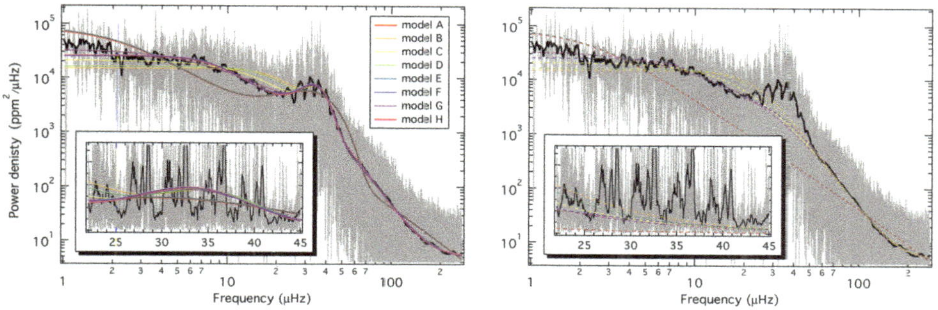

Figure 6.6. *Kepler* data (gray) of the KIC 7949599 power spectrum. Plotted in black is a heavily smoothed spectrum, and plotted in color are various background models with (left) and without (right) the Gaussian component. The insets zoom in on the area of the pulsational power excess. A total of eight models (labeled A through H) were tested. Adapted from Kallinger et al. (2014). Credit: Kallinger et al., A&A, 570, A41, 2014, reproduced with permission © ESO.

This particular formulation of background noise, due to Kallinger et al. (2014), is just one of the many versions proposed by numerous authors (e.g., Huber et al. 2009; Hekker et al. 2010; Karoff et al. 2013). The idea is then to fit the parameters of the background noise to the power spectrum of real stars that exhibit convection alongside *p*-mode pulsations. Figure 6.6 depicts the power spectrum of the red giant, KIC 7949599, along with eight different fits to the background, with (left) and without (right) the Gaussian component to account for the power excess due to pulsations (Kallinger et al. 2014). Thus, granulation amplitude σ_i, timescale τ_i, and slope α_i can be estimated by fitting the background of the power spectrum for any star that exhibits convection. Figure 6.7 depicts the distributions of these three parameters for 20,000+ *Kepler* targets (de Assis Peralta et al. 2017, 2018). This is one of the fundamental, well-established techniques in asteroseismology (Mathur et al. 2011; Karoff et al. 2013; Cranmer et al. 2014; Derekas et al. 2017).

The exceedingly difficult problem is proper description of convective energy transport. As convection is a macroscopic motion, energy transport is described by the 3D radiative magnetohydrodynamic (RMHD) equations employed over a large range of lengths (which, in turn, span large ranges of thermodynamic quantities—pressure, density, temperature) and over a large range of timescales. Numerical solutions to this problem exist (ANTARES, Muthsam et al. 2010; CO^{5}BOLD, Freytag et al. 2012) but are very time-consuming and come with a few caveats. Simplified treatments also exist (most notably the mixing-length theory; Böhm-Vitense 1958), but they fail to predict granulation and other convective effects near and at the surface of the star.

Convection in stellar atmospheres encompasses an impressive range of celestial objects: from the coolest *M* dwarfs to the F-type main-sequence stars, all red giants, and white dwarfs up to ~15,000 K (Tremblay et al. 2013). Even more impressive is the fact that the bulk properties of convection seem to scale particularly well among these diverse objects. Figure 6.8 depicts four examples from Tremblay et al. (2013), simulated by the CO^{5}BOLD code: a K giant, a K main-sequence star, a G main-

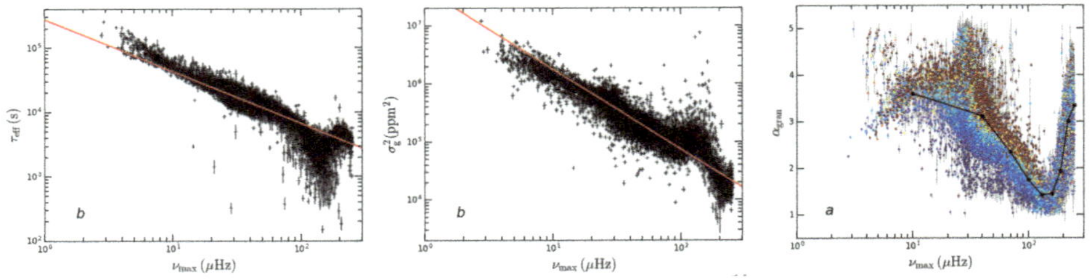

Figure 6.7. Granulation properties of 20,000 *Kepler* objects derived by fitting the background model to the power spectra. Left: effective timescales. Middle: characteristic amplitudes. Right: granulation slopes. The red line in the left and middle panels represents the best-fit scaling relation, while color-coding in the right panel corresponds to *Kepler* magnitude, red being the brightest and blue being the faintest objects. Credit: From R. de Assis Peralta, et al. "A new method for extracting seismic indices and granulation parameters: Results for more than 20,000 CoRoT and Kepler red giants" 339. 2-3, Copyright © (2018) by John Wiley Sons, Inc. Reprinted by permission of John Wiley & Sons, Inc.

sequence star, and a 100% hydrogen white dwarf (left to right, respectively). Atmospheric parameters and the intensity contrast are given above the panels. The similarity is remarkable given the change in scale—a whopping six orders of magnitude.

This observation demonstrates that it is possible to relate photospheric parameters (effective temperature, surface gravity, metallicity) to convection parameters (characteristic size, timescale, and intensity contrast). To do that, note that the pressure scale height, $H = p/\rho g$, is inversely proportional to surface gravity g and—provided that we neglect ∇_μ—directly proportional to effective temperature as $p \propto T_{\mathrm{eff}}$. Because the pressure scale height determines the characteristic size of a convection cell, therein lies the sought connection with atmospheric parameters. Similarly, characteristic lifetime can be estimated as $\tau = H/v_{\mathrm{rms}}$, where v_{rms} is convective velocity:

$$v_{\mathrm{rms}}^2 = \langle v \rangle^2 - \frac{\langle \rho v_x \rangle^2 + \langle \rho v_y \rangle^2 + \langle \rho v_z \rangle^2}{\langle \rho \rangle^2}. \tag{6.21}$$

Tremblay et al. (2013) provide empirical relations that connect characteristic size and decay time with atmospheric parameters:

$$D\ [\mathrm{km}] = 13.5 g^{-1} (T_{\mathrm{eff}} - 300 \log g)^{1.75}\, 10^{0.05[\mathrm{Fe/H}]}, \tag{6.22}$$

$$\tau\ [\mathrm{s}] = 2.08 g^{-1} (T_{\mathrm{eff}} - 300 \log g)^{1.75}\, 10^{0.05[\mathrm{Fe/H}]}. \tag{6.23}$$

Here T_{eff} is given in K, $\log g$ in cgs units, and [Fe/H] in dex.

Reality is, of course, much more complex. Cool stars feature brightness fluctuations that are not due to any one phenomenon, but a combination of phenomena: gravity darkening due to stellar rotation and tidal deformation, granulation, acoustic oscillations, and other chromospheric activity (spots, filaments, plages,

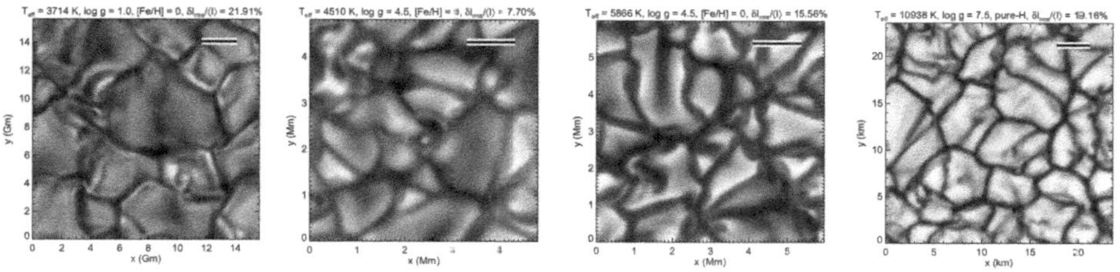

Figure 6.8. Modeled emergent bolometric intensity at the top of the horizontal xy plane for (left to right) a giant, a K-type main-sequence star, a G-type main-sequence star, and a hydrogen white dwarf. Atmospheric parameters, along with the intensity change, are listed above the plots. Note the scale change across the plots. The length of the upper bar on the top right of each plot corresponds to 10× scale height at optical depth 1. The models are computed using CO^{5}BOLD. Adapted from Tremblay et al. (2013). Credit: Tremblay et al., A&A, 557, 1, 2013, reproduced with permission © ESO.

spicules, protuberances, etc.; see Figure 6.9). These effects make the appearance of the stellar surface particularly complex and difficult to interpret, especially if that stellar surface is unresolved. The saving grace that allows us to disentangle granulation from chromospheric "blemishes" (Galilei 1613) is the difference in spatial distribution and associated timescales of these phenomena. Granulation is largely uniform across the entire star and is predominantly exhibited as anomalous limb darkening, while other phenomena are localized and cause bumps and wiggles in light curves. Ultimately, though, a model is just that, a model, and it is bound to fail in some circumstances, such as heavily spotted low-mass dwarfs.

6.3.1 Convective Blueshift

As demonstrated by Equations (6.22) and (6.23) and by Figure 6.8, the number of granules varies significantly over spectral types and luminosity classes. It is tempting to propose eclipsing binary stars as instruments to characterize convection of the eclipsed star, but this turns out to be beyond the current observing capabilities. For the Sun we expect $\sim 10^6$ granules, so the transit of Jupiter that blocks $\sim 1\%$ of the area covers $\sim 10^4$ granules—too many to draw any spatial inferences, as the signal of 10^4 granules will average out. The situation improves if we consider giant–main-sequence pairs or even giant–white-dwarf pairs, but other complicating factors such as spots, ficules, plages, and pulsations hinder our ability to directly probe the spatial characteristics of granulation.

Yet there is a tracer that we can use to detect convection: *convective blueshift*. As convective cells churn, the hotter, brighter gas is rising and the cooler, fainter gas is sinking. This causes a net blueshift in the radial direction (or, more precisely, along surface normals) that depends on the intensity contrast, characteristic size, and velocity profile of the granule. Assuming convection uniformity across the star, this net blueshift corresponds to the effective velocity $v_{\rm cb}$, which is always negative and causes the shifting of and the asymmetry in spectral lines. For the Sun, $v_{\rm cb} = -300$ m s^{-1}, and it ranges between -200 m s^{-1} for K-type main-sequence stars and -1000 m s^{-1} for F-type main-sequence stars (Shporer & Brown 2011).

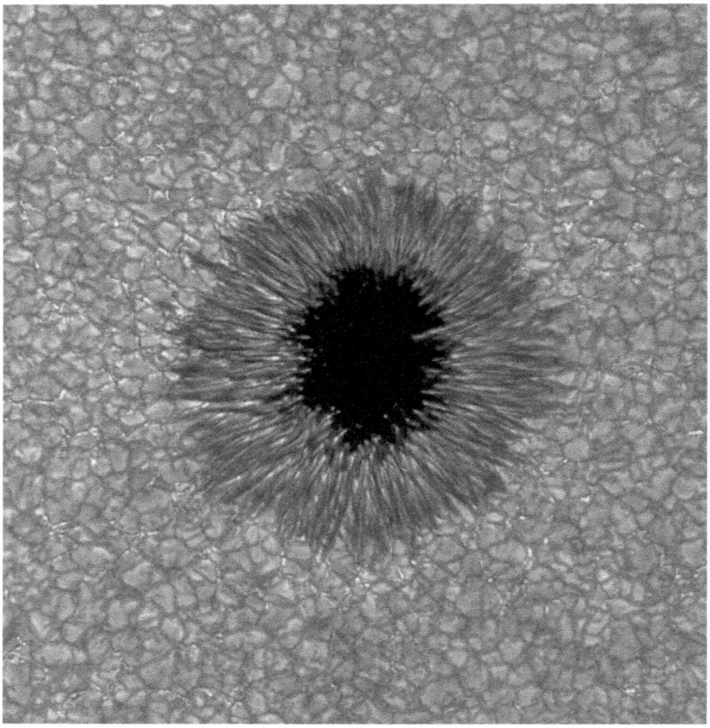

Figure 6.9. Magnified area around a spot on the Sun, where granulation surrounding it is still clearly visible. Image taken in the TiO band. Credit: Big Bear Solar Observatory/New Jersey Institute of Technology.

This effect is certainly observable during eclipses and transits as it modulates the velocity field of the stellar surface. Because the direction of v_{cb} is normal to the surface, photometrically this represents an "anomalous limb-darkening" effect, while spectroscopically it represents an "anomalous Rossiter–McLaughlin" effect. While fundamentally distinct from both limb darkening and the Rossiter–McLaughlin effects, convective blueshift causes the same types of signatures: photometrically via Doppler boosting and spectroscopically by shifting spectral lines and making them asymmetric. While the photometric signature is of the order of $\sim$10 ppm, i.e., under the detection threshold of modern observations, the radial velocity signature of tens to hundreds of m s^{-1} is certainly measurable. Figure 6.10 demonstrates this for the case of a slowly rotating Sun-like star being eclipsed by a smaller star. It is important to stress here that neglecting the convective blueshift effect can lead to anomalous measurements of the Rossiter–McLaughlin effect and, consequently, an incorrect determination of misalignment angles (see Section 6.1).

The future might be bright for the studies of convection. 3D RMHD models are believed to reproduce the circumstances on stellar surfaces to a sufficient accuracy, and planetary transits are proposed as probes of granulation (Dravins et al.

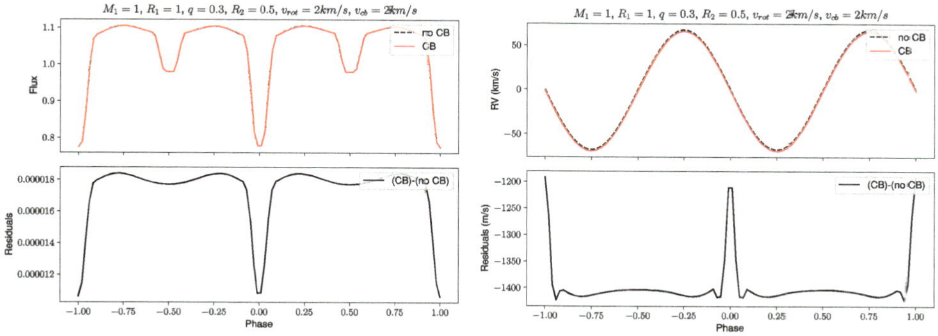

Figure 6.10. Effect of convective blueshift. Left: the effect on light curves is of the order of ~10 ppm, i.e., below the detection threshold of modern observations. Right: the effect on the radial velocity curve shows in the shift and the shape modulation during eclipses, both at levels easily observable with modern spectrographs.

2017a, 2017b) by way of time-resolved differential spectroscopy and the study of spectral line profiles.

6.4 Color Indices and Effective Temperatures

One of the main difficulties of modeling eclipsing binaries is accurate determination of individual temperatures of both components. Frequent practice in the literature is to *assume* the temperature of one star (i.e., from spectra or color indices) and fit the temperature of the other star. This approach is often inadequate, particularly for binaries with similar component temperatures and luminosities: in such cases, the contribution of both components to the system luminosity is significant, and it is difficult to accurately estimate the contribution of only one star in advance.

Before we describe a method capable of providing individual temperatures from standard photometry observations without any a priori assumptions, it proves useful to introduce a concept of effective temperature *of the binary*. A binary may be regarded as a point source, the effective temperature of which varies in time. Both components contribute to this effective temperature according to their projected sizes (including any eclipse events) and individual temperatures. Such a quantity is directly related to the color index, so it is observable.

The scaling of observed flux is typically done relative to another source, and the scaling of flux in different passbands is typically done individually, without any constraint between the scaling constants. In reality, these scaling constants are *not* independent; they are related to one another by the integrated composite spectra in a given passband. In other words, the scaling for each passband depends on passband luminosity, given by Equation (4.17), and the ratios of individual passband luminosities constrain the scaling of observed fluxes. If this is not taken into account, all color information contained in the observed light curves is lost. In consequence, the information on effective temperature of the binary is lost as well.

In the past two decades observers have put in substantial effort to scan the sky for standard stars to be used for photometric calibration: Landolt (1992) along the celestial equator, Landolt (2007) for −50° declinations, Landolt (2013) for +50°

declinations, Henden & Honeycutt (1997) and Bryja & Sandtorf (1999) covering fields around cataclysmic variables, Henden & Munari (2000) covering fields around symbiotic binaries, etc. These efforts help overcome the problem of small CCD fields with few or no photometric standards existing for calibration. Once standard stars are available in the field, they can be used to do all-sky (absolute) photometry of eclipsing binary targets. This effectively means that using measured color indices as additional information is plausible even if the data were not obtained under photometric conditions.

Here is a simple recipe that follows Prša & Zwitter (2006) on how to make sure that color information is retained. We can regard passband luminosities as simple level-setting quantities—physical context comes in only after we place constraints on their ratios based on the observed color indices. We should use a single, passband-independent parameter m_0 to transform *all* light curves from observed standard magnitudes to fluxes. The value of m_0 should be chosen so that the out-of-eclipse fluxes of the faintest light curve are of the order of unity. This m_0 is a single quantity for all light curves, which immediately implies that the magnitude difference, now the flux ratio, is preserved; hence, the color index is preserved. If the distance to the binary is known (e.g., from astrometry), m_0 immediately yields *observed* passband luminosities in physical units; intrinsic luminosities in physical units are obtained if the color excess $E(B - \mathrm{V})$ is also known.

This is where physics comes in: from such a set of observations, the ratios of passband luminosities provide us with the constraints we need—their ratios connect the theoretical quantity that depends on the temperature (effective temperature of the binary) with the observed quantity (the color index). Given that relation, we can infer the effective temperature of the binary from the color–temperature calibration. One such option is to use the updated Flower (1996) tables with corrected coefficients given in Table 6.3.

Applying the color constraint, effective temperatures of individual components may be readily disentangled by the minimization method. The method is now able to

Table 6.3. Coefficients of the Empirical log T_{eff}–$(B - V)$ Calibration Formula Given by the Seventh-degree Polynomial Fit $\log T_{\mathrm{eff}} = \sum_{i=0}^{7} C_i(B - V)^i$ (Flower 2004, private communication)

Coefficient	Lum. Class V, IV, III, II	Lum. Class I
C_0	3.979145	4.012560
C_1	−0.654992	−1.055043
C_2	1.740690	2.133395
C_3	−4.608815	−2.459770
C_4	6.792600	1.349424
C_5	−5.396910	−0.283943
C_6	2.192970	–
C_7	−0.359496	–

Note. The second column applies to main-sequence stars, subgiants, and giants; the third column applies to supergiants.

find only those combinations of parameters that satisfy the constraints imposed on the effective temperature of the binary. As the relationship between effective temperatures of individual components is fully determined by the light-curve shape (predominantly by the eclipse depth ratios), and given that the sum of both component contributions must match the effective temperature of the binary, the color-constrained minimization yields effective temperatures of individual components without any a priori assumptions.

To demonstrate the importance of color constraining, consider two stars of similar luminosities but different effective temperatures, say, F8V and G1V, that we observe in two photometric passbands, say, B and V (for the exact details of the simulation, see Prša & Zwitter 2006). We can then synthesize their color index, say, $B - V$, as a function of phase. Using the coefficients in Table 6.3, we can convert $B - V$ to effective temperature of the binary, also as a function of phase. Now we contrast three approaches: (1) we do not make any inferences on the temperatures from the data; (2) we derive the effective temperature from the color index, assign it to the primary star, and fit the temperature of the secondary star, and (3) we apply the color index constraint by imposing the passband luminosity ratio at a certain (yet arbitrary) out-of-eclipse phase to agree with observations.

We expect that, in the first case, we obtain a family of solutions where the values of $T_{\mathrm{eff}1}$ and $T_{\mathrm{eff}2}$ are strongly correlated, and we cannot discern which combination of temperatures is correct. In the second case we expect that this degeneracy is broken, but it is broken *incorrectly*, as we impose the value of $T_{\mathrm{eff}1}$ that is not correct. In the third case we expect that the degeneracy is broken again, but this time yielding correct $T_{\mathrm{eff}1}$ and $T_{\mathrm{eff}2}$. Figure 6.11 depicts a $T_{\mathrm{eff}1}$–$T_{\mathrm{eff}2}$ cross section of the parameter space, where all other parameters are adjusted at every point. It showcases the severe correlation of individual temperatures when the color index constraint is not applied. The reason for this is the degeneracy between the temperatures and passband luminosities: local effects of slightly different temperatures (limb darkening, reflection, etc.) on the shape of the light curve are very small, so the cost

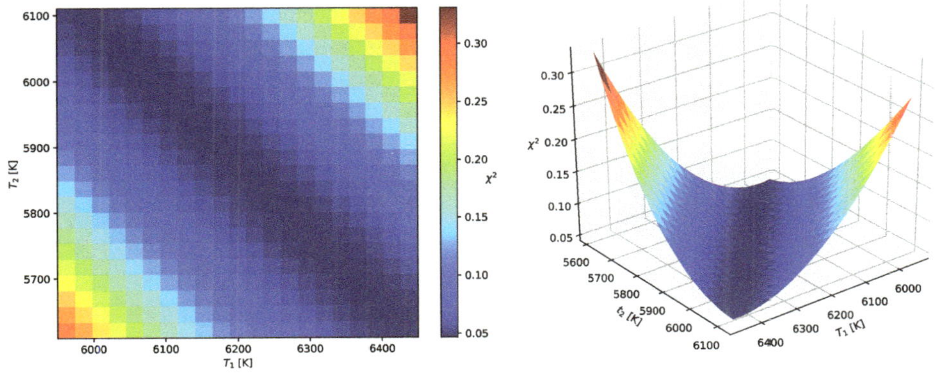

Figure 6.11. $T_{\mathrm{eff}1}$–$T_{\mathrm{eff}2}$ cross section of the parameter hyperspace. The color scheme denotes the value of the cost function (sum of squares of the residuals): black corresponds to the lowest cost function value, and red corresponds to the highest cost function value.

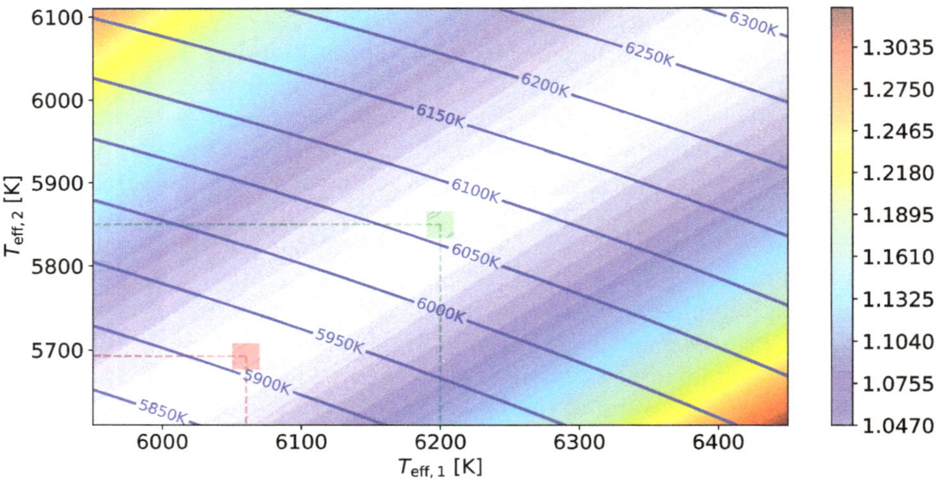

Figure 6.12. Contours of the $T_{\mathrm{eff}1}$–$T_{\mathrm{eff}2}$ cross section of the parameter space overplotted by the contours of constant effective temperature of the binary. The inferred effective temperature of the binary from the $B - V$ color index is $T_{\mathrm{eff}} = 6060$ K. The green square corresponds to the color-constrained value, and the red square corresponds to the value obtained by setting $T_{\mathrm{eff}1} \equiv T_{\mathrm{eff}}$. The systematic error in this case is ~140 K, while the color-constrained solution is within 25 K of the correct solution in both individual temperatures.

function is largely insensitive to their variation. However, even slightly different temperatures cause an additional effect that is an order of magnitude larger: the scaling of the color index. If this change is not constrained, the model relies only on the aforementioned second-order effects that are frequently buried in data noise and parameter degeneracy. Without color constraining, there is thus no practical way to pinpoint a location in the $T_{\mathrm{eff}1}$–$T_{\mathrm{eff}2}$ cross section that corresponds to the right solution. Figure 6.12 shows the impact of color constraining. Color contours represent the same parameter space cross section as the one depicted in Figure 6.11; plotted in blue are the lines of constant effective temperature of the binary. For this particular case, the value that follows from the $B - V$ color index is $T_{\mathrm{eff}} = 6060$ K. The red square corresponds to case 2, where we assign T_{eff} to the individual temperature of the primary. That breaks the degeneracy; however, the assigned temperature is ~140 K too low, and the derived temperature of the secondary star is ~150 K too low. This systematic error is avoidable, though; the green square corresponds to case 3, where color constraining was applied. In this case both temperatures are within ~25 K of their correct values. Unless the luminosity of one component dominates the system (i.e., contributes 99% or more of total light), assigning effective temperature of the binary to one star will always result in systematic error greater than the uncertainty in individual temperatures.

This demonstrates how severe the implications of subjectively assuming a primary star temperature from the color index may be if special care is not taken to reiterate the acquired solution until the calculated and observed color indices do not match. The effect is already very pronounced for a modest case of two similar main-sequence components, and it would be significantly greater for a hot main-sequence–

cool giant system, where both stars have similar luminosities. In those cases the published solution may be off by as much as ~1000 K or more.

6.5 Limitations

The current state of our understanding of the underlying physics processes is reasonable, but far from perfect. To end this chapter, we briefly mention several interesting directions that have yet to witness an in-depth study.

DIFFERENTIAL ROTATION OF ECLIPSING BINARY COMPONENTS. This is an area of active research, and our understanding and ability to deduce differential rotation from observations are still limited. There have been several detection reports in the literature, i.e., HR 1099 (Petit et al. 2001), PV Cas (Yıldız 2005), V471 Tau (Hussain et al. 2006), and HD 155555 (Dunstone et al. 2008), but these results still seem preliminary for a systematic treatment in modeling codes.

DETAILED PHYSICS OF ACCRETION DISKS AND HOT SPOTS. Many semi-detached, cataclysmic, and symbiotic systems show clear signs of mass accretion from the lobe-filling component to the primary star. Several tools have the capability to model such disks, most notably BINSYN (Linnell et al. 2007), ELC (Orosz & Hauschildt 2000), and the code by Zhang et al. (1986), but it has become apparent that detailed treatment of the processes is necessary, through either particle simulations (Kreidberg et al. 2010) or magnetohydrodynamics (Bisikalo 2010). A toy model would be the only feasible approach for including this effect, but given the associated complexities, this might not result in a profound understanding of the underlying physics.

TRANSIENT PHENOMENA. Sun-like variability on a ~30–50 ppm level, complex spot structures, flares, and similar phenomena are systematically seen in *Kepler* data. However, most of those transient phenomena may be treated as perturbations and do not significantly affect other binarity effects. While the addition of intrinsic and tidally induced pulsations to the model is a feasible goal, the current state of binary star asteroseismology is still maturing. The reports on "heartbeat stars," eccentric ellipsoidal binaries that exhibit notable periastron brightening and common tidally induced pulsations (Hambleton et al. 2013; Thompson et al. 2012), sound like a very promising laboratory for measuring internal structure factors.

POPULATION AND EVOLUTION THEORY AS DEGENERACY BREAKERS. Several tools, e.g., DEBiL (Devor 2005) and BLENDER (Torres et al. 2011), use constraining methods to help break the degeneracy. In particular, the components in a binary are assumed to be coeval, and their physical properties are deduced from stellar evolution models. At other times, M–L–R–T calibrations across the H-R diagram may be used to constrain the values of the derived parameters. Both DEBiL and BLENDER have a specific purpose, substantially different from the general modeling effort, and that is to estimate a set of plausible configurations that would reproduce a given data set. For state-of-the-art modeling engines, however, this poses an inherent risk of circular arguments, as eclipsing binary results are predominantly used to calibrate ages and H-R diagram relationships.

References

Abdul-Masih, M., Prša, A., Conroy, K., et al. 2016, AJ, 151, 101
Albrecht, S., Winn, J. N., Butler, R. P., et al. 2012, ApJ, 744, 189
Albrecht, S., Winn, J. N., Torres, G., et al. 2014, ApJ, 785, 83
Anderson, D. R., Hellier, C., Gillon, M., et al. 2010, ApJ, 709, 159
Astronomical Almanac 2017, Astronomical Almanac for the Year (Washington, DC: US Gov Printing Office)
Avni, Y., & Schiller, N. 1982, ApJ, 257, 703
Bahng, J., & Schwarzschild, M. 1961, ApJ, 134, 312
Bisikalo, D. 2010, in ASP Conf. Ser. 435, Binaries—Key to Comprehension of the Universe ed. A. Prša, & M. Zejda (San Francisco, CA: ASP), 287
Bloemen, S., Marsh, T. R., Degroote, P., et al. 2012, MNRAS, 422, 2600
Böhm-Vitense, E. 1958, ZA, 46, 108
Borkovits, T., Érdi, B., Forgács-Dajka, E., & Kovács, T. 2003, A&A, 398, 1091
Borkovits, T., Forgács-Dajka, E., & Regály, Z. 2007, A&A, 473, 191
Borkovits, T., Forgács-Dajka, E., & Regály, Z. 2011, A&A, 528, A53
Borucki, W. J., Koch, D., Basri, G., et al. 2010, Sci, 327, 977
Bryja, C., & Sandtorf, J. R. 1999, AAS Meeting, 194, 115.03
Chiavassa, A., Haubois, X., Young, J. S., et al. 2010, A&A, 515, A12
Conroy, K. E., Prša, A., Horvat, M., & Stassun, K. G. 2018, arXiv:1801.01063 [astro-ph.SR]
Conroy, K. E., Prša, A., Stassun, K. G., et al. 2014, AJ, 147, 45
Cranmer, S. R., Bastien, F. A., Stassun, K. G., & Saar, S. H. 2014, ApJ, 781, 124
Dawson, R. I., & Chiang, E. 2014, Sci, 346, 212
de Assis Peralta, R., Samadi, R., & Michel, E. 2018, AN, 339, 134
de Assis Peralta, R., Samadi, R., & Michel, É. 2017, in EPJ Web of Conf. 160, Seismology of the Sun and the Distant Stars—Using Today's Successes to Prepare the Future, ed. M. J. P. F. G. Monteiro, M. S. Cunha, & J. M. T. S. Ferreira (Les Ulis: EDP Science), 01012
Derekas, A., Plachy, E., Molnár, L., et al. 2017, MNRAS, 464, 1553
Devor, J. 2005, ApJ, 628, 411
Doyle, L. R., Carter, J. A., Fabrycky, D. C., et al. 2011, Sci, 333, 1602
Dravins, D., Ludwig, H.-G., Dahlén, E., & Pazira, H. 2017a, A&A, 605, A90
Dravins, D., Ludwig, H.-G., Dahlén, E., & Pazira, H. 2017b, A&A, 605, A91
Dunstone, N. J., Hussain, G. A. J., Collier Cameron, A., et al. 2008, MNRAS, 387, 1525
Flower, P. J. 1996, ApJ, 469, 355
Freytag, B., Steffen, M., Ludwig, H. G., et al. 2012, JCoPh, 231, 919
Galilei, G. 1613, Istoria e dimostrazioni intorno alle macchie solari e loro ac-cidenti comprese in 3 lettere (Roma: Giacomo Mascardi)
Hambleton, K. M., Kurtz, D. W., Prša, A., et al. 2013, MNRAS, 434, 925
Harvey, J. 1985, in Proc. ESA Workshop, Vol. 235, Future Missions in Solar, Heliospheric & Space Plasma Physics, ed. E. Rolfe, & B. Battrick (Netherlands: ESA), 199
Hekker, S., Broomhall, A.-M., Chaplin, W. J., et al. 2010, MNRAS, 402, 2049
Henden, A., & Munari, U. 2000, A&AS, 143, 343
Henden, A. A., & Honeycutt, R. K. 1997, PASP, 109, 441
Herschel, W. 1801, RSPT, 91, 265
Hirano, T., Narita, N., Sato, B., et al. 2011, PASJ, 63, L57

Horvat, M., Conroy, K. E., Pablo, H., et al. 2018, ApJS, 237, 26
Huber, D. 2015, in Giants of Eclipse: The ζ Aurigae Stars and Other Binary Systems, Astrophysics and Space Science Library, Vol. 408 (Basel: Springer), 169
Huber, D., Stello, D., Bedding, T. R., et al. 2009, CoAst, 160, 74
Hussain, G. A. J., Allende Prieto, C., Saar, S. H., & Still, M. 2006, MNRAS, 367, 1699
Irwin, J. B. 1952, ApJ, 116, 211
Kallinger, T., De Ridder, J., Hekker, S., et al. 2014, A&A, 570, A41
Kallrath, J., & Milone, E. F. 2009, Eclipsing Binary Stars: Modeling and Analysis (New York: Springer)
Karoff, C., Campante, T. L., Ballot, J., et al. 2013, ApJ, 767, 34
Kippenhahn, R., & Weigert, A. 1994, Stellar Structure and Evolution (Basel: Springer)
Kirk, B., Conroy, K., & Prša, A. 2016, AJ, 151, 68
Kreidberg, L., Wood, M., & Wetzstein, M. 2010, AAS Meeting, 215, 268
Kruszewski, A. 1967, in On the Evolution of Double Stars, Vol. 17, ed. J. Dommanget (Cambridge: Cambridge Univ. Press), 146
Landolt, A. U. 1992, AJ, 104, 340
Landolt, A. U. 2007, AJ, 133, 2502
Landolt, A. U. 2013, AJ, 146, 131
Limber, D. N. 1963, ApJ, 138, 1112
Linnell, A. P., Godon, P., Hubeny, I., Sion, E. M., & Szkody, P. 2007, ApJ, 662, 1204
Mathur, S., Hekker, S., Trampedach, R., et al. 2011, ApJ, 741, 119
Mayor, M., Pepe, F., Queloz, D., et al. 2003, Msngr, 114, 20
Mazeh, T., Perets, H. B., McQuillan, A., & Goldstein, E. S. 2015, ApJ, 801, 3
Muthsam, H. J., Kupka, F., Löw-Baselli, B., et al. 2010, NewA, 15, 460
Orosz, J. A. 2015, in ASP Conf. Proc. 496, Living Together: Planets, Host Stars and Binaries, ed. S. M. Rucinski G. Torres, & M. Zejda (San Francisco, CA: ASP), 55
Orosz, J. A., & Hauschildt, P. H. 2000, A&A, 364, 265
Orosz, J. A., Welsh, W. F., Carter, J. A., et al. 2012a, ApJ, 758, 87
Orosz, J. A., Welsh, W. F., Carter, J. A., et al. 2012b, Sci, 337, 1511
Petit, P., Donati, J.-F., Wade, G. A., et al. 2001, in Springer Lecture Notes in Physics, Vol. 573, Astrotomography, Indirect Imaging Methods in Observational Astronomy, ed. H. M. J. Boffin, D. Steeghs, & J. Cuypers (Berlin: Springer), 232
Philippov, A. A., & Rafikov, R. R. 2013, ApJ, 768, 112
Pols, O. R. 2011, Stellar Structure and Evolution (Basel: Springer)
Prusti, T., de Bruijne, J. H. J., Brown, A. G. A., et al. 2016, A&A, 595, A1
Prša, A., Batalha, N., Slawson, R. W., et al. 2011, AJ, 141, 83
Prša, A., & Zwitter, T. 2006, Ap&SS, 304, 347
Rappaport, S., Deck, K., Levine, A., et al. 2013, ApJ, 768, 33
Rappaport, S., Nelson, L., Levine, A., et al. 2015, ApJ, 803, 82
Shporer, A., & Brown, T. 2011, ApJ, 733, 30
Sousa, S. G., Santos, N. C., Israelian, G., Mayor, M., & Udry, S. 2011, A&A, 533, A141
Thompson, S. E., Everett, M., Mullally, F., et al. 2012, ApJ, 753, 86
Torres, G., Fressin, F., Batalha, N. M., et al. 2011, ApJ, 727, 24
Tremblay, P. E., Ludwig, H. G., Freytag, B., Steffen, M., & Caffau, E. 2013, A&A, 557, A7
Triaud, A. H. M. J. 2017, arXiv:1709.06376 [astro-ph.EP]

Welsh, W. F., Orosz, J. A., Carter, J. A., et al. 2012, Natur, 481, 475
Winn, J. N., & Fabrycky, D. C. 2015, ARA&A, 53, 409
Winn, J. N., Johnson, J. A., Marcy, G. W., et al. 2006, ApJL, 653, L69
Yıldız, M. 2005, MNRAS, 363, 967
Zhang, E. H., Robinson, E. L., & Nather, R. E. 1986, ApJ, 305, 740

Part III

Modeling Considerations

Chapter 7

Numerical Model

So far we have laid down the theoretical framework that describes our current understanding of the physics governing the dynamics and radiative properties of binary (and multiple) stars. In this section we focus our attention on (mostly numerical) considerations that are used to approximate radiating surfaces. We should emphasize from the get-go that the methodology presented here represents one possible framework, and it should be in no way construed as the only one. Other techniques have been employed across the field; Kallrath and Milone (2009) provide an overview of the most important ones. As is the case in the rest of the book, we do not aim to be complete, but instead we seek to provide a concise review of what we consider best practices.

In a nutshell, generating a synthetic data set (light curves, radial velocity curves, astrometric positions, and other observables) broadly consists of the following sequence of steps:

SURFACE DISCRETIZATION: represent the surfaces of orbiting bodies with a discrete mesh: a set of surface elements that cover the surface;

ORBITAL DYNAMICS: given either the analytical framework (for Keplerian systems) or numerical integration (for interacting, dynamical systems), position and orient the objects in space as a function of time;

COMPUTATION OF LOCAL INTENSITY: for each surface element, compute local intensity in the direction perpendicular to the surface; intensity depends on temperature, surface gravity, heavy metal abundance, tidal and rotational distortion, spots, etc.;

PLANE-OF-SKY PROJECTION: compute the placement of objects with respect to the observer and determine view obstruction (horizon, eclipses);

FLUX COMPUTATION: compute local limb-darkened intensity in the direction of the observer and integrate over the visible surfaces to obtain a desired synthetic observable.

7.1 Discretization of Stellar Surfaces

Computing the amount of flux received from an EB becomes a nontrivial task as soon as there is even a marginal surface distortion of the EB components. The employed generalized Roche geometry provides us with that distortion, and a range of related phenomena are then taken into account: surface brightness variation due to gravity darkening, limb darkening and reflection, ellipsoidal variations due to the changing cross-section size that faces the observer, spots coming in and out of view, etc. The dynamical aspects then render the two components onto their respective orbits, and numerical integration is done over the visible surfaces, determined by eclipses and visible cross sections, to compute total flux. It is thus imperative that local conditions across the surfaces of both components be approximated as well as possible.

That said, we also need to consider the potential gain in model accuracy and evaluate it with respect to its computational cost. Having rigorous expressions built into the model to describe smooth variations in parameter values across stellar surfaces would be ideal in terms of accuracy, but at the same time they would render such codes useless because of the excessive time required for the computation. On the other hand, having an overly simplified geometry or a set of assumptions, e.g., a predetermined distribution of physical quantities over such surfaces, would be ideal in terms of speed, but at a severe expense of accuracy. We seek the middle grounds: without making any prior assumptions about the shape of the stars or parameter variations over their surfaces, we build the model on the formalism of Roche equipotentials (Section 3.2) and *sample* parameter values across the surface in a discrete set of surface elements uniformly distributed along the stellar surface. The underlying physics is there in all its analytical glory, but it is used only in these discrete surface elements. The number of elements that are required to reach a given level of accuracy depends on the problem at hand: if a modeled system is well detached, then a limited number of surface elements are sufficient for accurate results. On the other hand, if a modeled system is in contact or features other complexities, then a number of surface elements should be noticeably larger.

7.1.1 Trapezoidal Discretization

Trapezoids have proven particularly convenient to cover the surfaces of slightly distorted stars. This technique is the basis of the widely used Wilson–Devinney code (WD; Wilson and Devinney 1971). WD is a publicly available code that underwent many updates, improvements, and fine-tuning (Wilson and Biermann 1976; Wilson and Sofia 1976; Wilson 1979, 1990; Milone et al. 1992; Kallrath et al. 1998; van Hamme and Wilson 2003; Prša and Zwitter 2005; Van Hamme and Wilson 2007; Wilson 2008; Wilson and Van Hamme 2014, to name just a few), which firmly established it as the model of choice for many eclipsing binary stars in the literature.

Points on the stellar surface that constitute trapezoidal centers are sampled uniformly in latitude and longitude. Recall from Section 3.1, Figure 3.2, that latitude is by definition measured from the $+z$-axis direction ($\theta = 0$) to the $-z$-axis direction

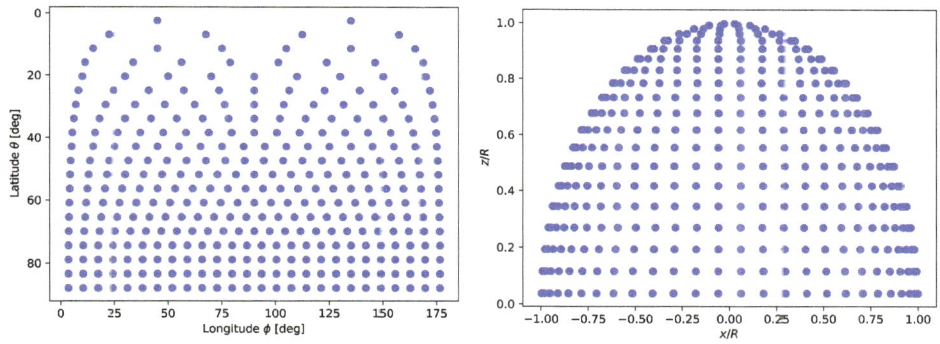

Figure 7.1. Sampling the stellar surface. Left: sampled points in latitude (θ) and longitude (ϕ) according to Equations (7.1) and (7.2) and $N = 20$. Right: same as the left panel, but converted into the xz plane.

($\theta = \pi$). For as long as we discuss aligned cases, stellar shapes are symmetric with respect to the equator; it then suffices to sample only one-half of the hemisphere (one-quarter of the star) and mirror the points to the other three-quarters. Given the number of points N, colatitude is divided into equidistant intervals:

$$\theta_k = \frac{\pi}{2}\frac{k - 0.5}{N}, \qquad k = 1 \ldots N. \tag{7.1}$$

Given θ_k, the longitudinal line at that θ_k is then divided into equidistant intervals:

$$\begin{aligned} M_k &= 1 + \text{int}\,(1.3N \sin\theta_k), \\ \phi_l &= \pi\frac{l - 0.5}{M_k}, \qquad l = 1 \ldots M_k. \end{aligned} \tag{7.2}$$

Longitude ϕ is by definition measured from the direction to the companion star in a counterclockwise direction (see Equations (3.9)). The factor 1.3 in Equation (7.2) means that 30% more points will be sampled in longitude than in latitude, which was traditionally introduced to improve the accuracy of partially eclipsed surface elements (Wilson 1993). Figure 7.1 shows the distributions of sampled points in latitude and longitude for $N = 20$.

This discretization provides us with surface element center points (ϕ_l, θ_k) and sizes $\Delta\phi_l \times \Delta\theta_k$. These are planar trapezoids, which is why this method is called trapezoidal discretization.

As the steps in colatitude and longitude are constant, this discretization scheme is very fast and surface elements are straightforward to compute. Note that, in the case of a misaligned binary, the reflection shortcut cannot be used and the number of surface elements in which we need to compute local conditions is quadrupled.

7.1.2 Cylindrical Trapezoidal Discretization

There are two notable deficiencies of the above-mentioned surface discretization: (1) as the distribution of surface elements is uniform in angles θ and ϕ, surface element size and positioning are nonuniform along equipotentials for distorted stars; and (2) as all surface elements begin at $\phi = 0$, there is a "seam" across $x = 0$ where the mesh is connected. Both deficiencies can be addressed by changing the coordinate system and revising the point sampling strategy.

We start by rewriting Equation (3.31) in cylindrical coordinates, $x = \varrho_\perp \cos\phi$, $y = \varrho_\perp \sin\phi$, $z = z$:

$$\Omega = \frac{1}{\sqrt{\varrho_\perp^2 + z^2}} + q\left(\frac{1}{\sqrt{\delta^2 + \varrho_\perp^2 + z^2 - 2\varrho_\perp\delta\cos\phi}} - \frac{\varrho_\perp\cos\phi}{\delta^2}\right) + \frac{1}{2}F^2(1+q)\varrho_\perp^2. \tag{7.3}$$

For any given value of z and ϕ, $\varrho_\perp$ may be computed by a Newton–Raphson method, analogous to Equation (3.36), where $z_{\text{pole}} \equiv \varrho_{\text{pole}}$:

$$f(\varrho_\perp) = \frac{1}{\sqrt{\varrho_\perp^2 + z^2}} + q\left(\frac{1}{\sqrt{\delta^2 + \varrho_\perp^2 + z^2 - 2\varrho_\perp\delta\cos\phi}} - \frac{\varrho_\perp\cos\phi}{\delta^2}\right) + \frac{1}{2}F^2(1+q)\varrho_\perp^2 - \frac{1}{z_{\text{pole}}} - \frac{q}{\sqrt{z_{\text{pole}}^2 + \delta^2}}; \tag{7.4}$$

$$f'(\varrho_\perp) = -\frac{\varrho_\perp}{(\varrho_\perp^2 + z^2)^{3/2}} - q\left(\frac{\varrho_\perp - \delta\cos\phi}{(\delta^2 + \varrho_\perp^2 + z^2 - 2\varrho_\perp\delta\cos\phi)^{3/2}} + \frac{\cos\phi}{\delta^2}\right) + F^2(1+q)\varrho_\perp; \tag{7.5}$$

$$\varrho_{\perp i+1} = \varrho_{\perp i} - \frac{f(\varrho_{\perp i})}{f'(\varrho_{\perp i})}. \tag{7.6}$$

Now we can slice the star as before (see Equations (7.1) and (7.2)), this time along z and ϕ (see Figure 7.2), adding one last detail: each successive band will be *offset* in ϕ by half the width of the surface element,

$$\phi_l = \pi\frac{l + [(l+1) \bmod 2]/2}{M_k}, \quad l = 1\ldots M_k, \tag{7.7}$$

where mod denotes the modulo, i.e., the remainder of integer division. Figure 7.3 depicts the results of this *zip*-scheme.

7.1.3 Surface Triangulation

Surface discretization approaches discussed thus far sample the surface uniformly in colatitude and longitude into planar, trapezoidal elements for which we assume

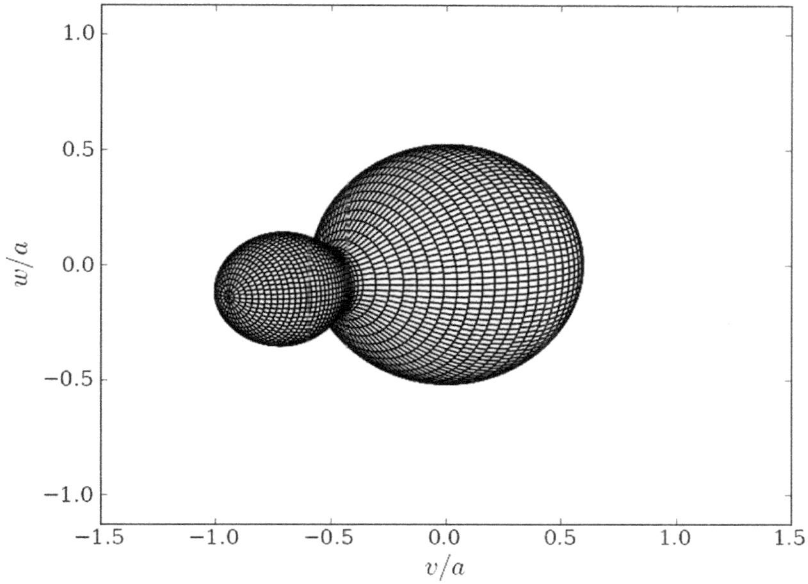

Figure 7.2. Cylindrical mesh of a contact binary. The neck area of the binary is much better sampled than with the original trapezoidal meshing.

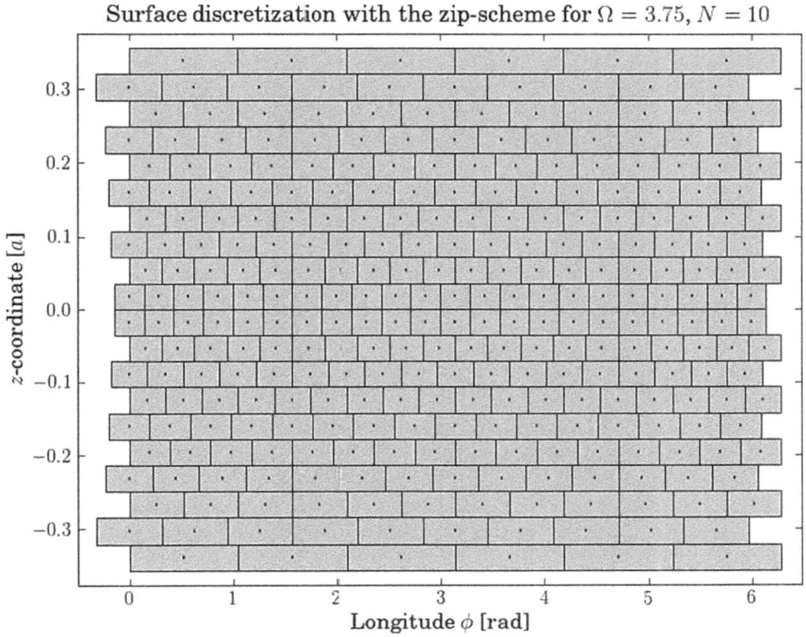

Figure 7.3. Sampling the stellar surface with the *zip*-scheme.

uniform local conditions. The most significant problem with this strategy is that the surface is approximated by a *disconnected, self-overlapping* mesh that does not cover the surface completely. The left panel in Figure 7.4 displays a part of the star near a pole, a region where these problems are particularly exacerbated. To combat this deficiency, surface element areas are not computed from the trapezoids themselves but from the discretized surface element areas, $\Delta\sigma = \Delta\theta\Delta\phi\rho^2 \sin\theta/(\cos\beta)^{-1}$, where ρ, θ, and ϕ are spherical coordinates, and β is the angle between the surface normal and the radius vector. The problems become more pronounced as components become distorted, and especially when we deal with ultraprecise photometric data, since a uniform sampling in colatitude and longitude implies that the sampling in terms of surface coverage is not uniform.

To overcome this problem, we can replace trapezoidal surface elements with triangles (Prša et al. 2016). The triangulation of implicit surfaces is a long-standing problem in computational physics. Hartmann (1998) presents a *marching method*: an algorithm that can discretize any implicit surface into a set of near-equilateral triangles, so that every element on the surface has approximately the same area, irrespective of the position or the amount of surface distortion. The right panel of Figure 7.4 demonstrates this: the polar region of the star has no inherent symmetry that would impose "seams" across the surface, or surface elements of notably different sizes. In effect, any systematics that may arise from surface discretization will be minimized. A mesh generated this way is computationally tractable, and it alleviates the problems with holes or overlaps between trapezoidal elements. An added benefit is that we can discretize any implicit geometric body (i.e., accretion disks, disintegrating planets, alien megastructures) and model any transient phenomena within the same formalism.

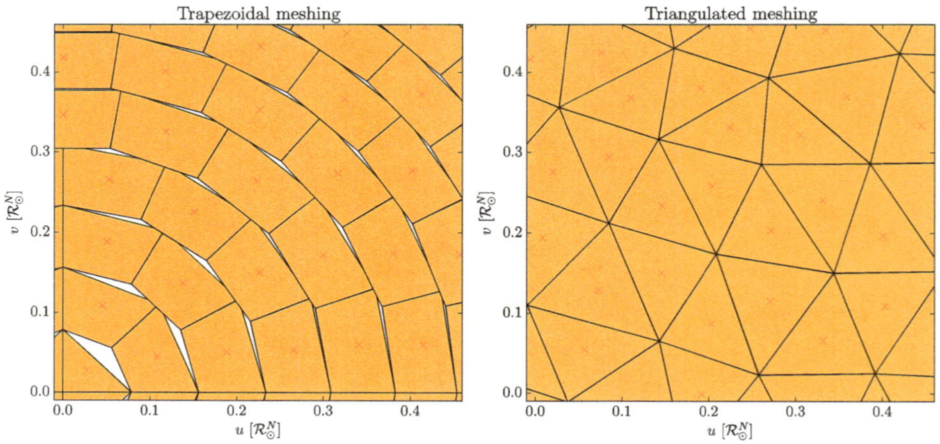

Figure 7.4. Mesh comparison between the trapezoidal discretization using trapezoidal elements (left) and triangulation (right). The crosses denote surface element centers. Both parts correspond to the same region on the star. The issues with trapezoidal discretization are disconnected, overlapping meshes with holes and obvious "seams" across the surface. All these cause systematic effects in computed fluxes. Adopted from Prša et al. (2016) © 2016. The American Astronomical Society.

The marching triangulation method discretizes a closed, three-dimensional parametric surface described by an implicit equation; in the case of binary stars, this is the effective surface potential, Equation (3.31). A mesh of triangles is built by the following scheme:

FIRST BLOCK. Generate the initial front polygon:

1. choose an initial point on the surface, $\boldsymbol{p}_0$;
2. determine the tangent plane to the surface in $\boldsymbol{p}_0$;
3. surround $\boldsymbol{p}_0$ with a regular hexagon; denote the vertices of this hexagon with $\boldsymbol{r}_1$ through $\boldsymbol{r}_6$;
4. project the vertices onto the surface to obtain surface points $\boldsymbol{p}_1$ through $\boldsymbol{p}_5$; points $\boldsymbol{p}_0$ through $\boldsymbol{p}_6$ form the first six triangles of the mesh.

The initial front polygon is the ordered array of points $\boldsymbol{p}_1, \dots, \boldsymbol{p}_6$.

SECOND BLOCK. Generate the rest of the mesh:

1. calculate the outer angle α_i at each point of the front polygon in the plane tangential to the surface in that point;
2. identify point $\boldsymbol{p}_i$ of the front polygon with the minimal angle α_i;
3. surround $\boldsymbol{p}_i$ by triangles with inner angles as close to 60° as possible;
4. remove point $\boldsymbol{p}_i$ from the front polygon and add the newly created triangle vertex/vertices to the front polygon.

Repeat the scheme above until the front polygon consists of only three points. That will be the last triangle of the mesh.

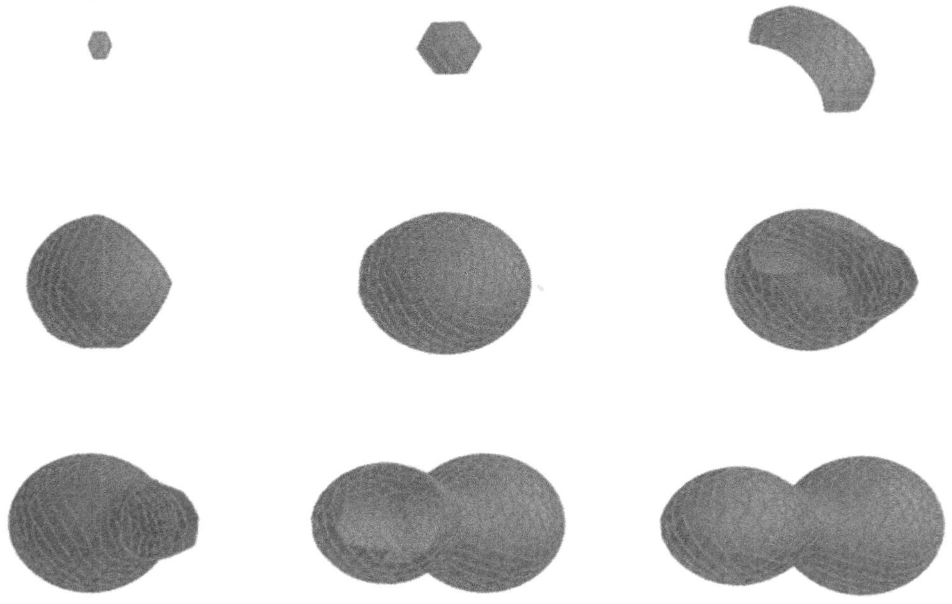

Figure 7.5. Mesh of triangles that approximates the surface of a contact binary. The parameters of the binary are $\Omega = 2.6$ and $q = 0.5$. The interactive version of the book animates the creation of the mesh. The animation is available online at https://doi.org/10.1088/978-0-7503-1287-5.

The scheme above is a simplified version of the more general triangulation method where the front polygon can be split into two or more polygons to allow for more complex topologies. Details can be found in Hartmann (1998). Figure 7.5 depicts the meshing process of a contact binary.

While triangulation does a much better job at generating equal surface element areas, the isosurface problem that the mesh approximates remains. While a trapezoidal mesh overestimates the area because surface element centers are on the isosurface, triangulation underestimates it because the vertices are on the isosurface. To correct for this, surface elements need to be offset such that (1) the total surface area of the mesh exactly equals the analytical surface area of the isosurface, and (2) each surface element approximates the local isosurface optimally, i.e., it is positioned in a way that exactly balances isosurface deviations due to curvature below and above the surface element. This is achieved by linearly shifting each vertex of the mesh, $\boldsymbol{p}_i \mapsto \boldsymbol{p}_i + \alpha_i \boldsymbol{n}_i$, $i = 1, \ldots, n$, where $\boldsymbol{n}_i = \nabla\Omega(r_i)/\|\nabla\Omega(r_i)\|$ is a normal to the isosurface in the ith vertex and α_i is the offset along the normal at that vertex.

Qualitatively we expect that shifts α_i have the same sign as the isosurface curvature and that their value should be larger for larger curvatures since larger curvatures imply larger distances to the neighboring triangles. Since the number of triangles is generally large, any local curvature across a single surface element can be considered constant. Thus, we can optimize α_i on a perfectly spherical mesh and apply those corrections to local surface elements. Sphere triangulation using the marching method results in a uniform set of triangles; in the limit of a large number of triangles, the optimal shift for all vertices is equal to $\alpha \sim A_t^2/4R$, where $1/R$ is the curvature of the sphere and A_t is the area of a single triangle. This result can be generalized to the expected functional form:

$$\alpha_i = \xi H_i S_i, \tag{7.8}$$

where ξ is a single scalar parameter for all shifts, H_i is the mean curvature of the underlying isosurface at the vertex approximated from the mesh (Taubin 1995), and S_i is one-third of the area of triangles sharing the ith vertex.

A significant difference between the marching and trapezoidal meshes arises in contact binary systems computed with the Roche model. The trapezoidal method tends to diverge near the neck of contact envelopes, which requires the meshing of the surface to account for each star separately and "glue" them together at the neck (Wilson 1979). This can often lead to poor coverage of the neck region, particularly for systems with a small mass ratio and large fillout factor. The marching method, on the other hand, computes the entire contact binary mesh as one surface, with complete coverage of the neck region. Examples of a trapezoidal and triangulated mesh for a contact system are given in Figure 7.6.

7.2 Dynamical Aspects

In Section 3.4 we discussed at length the basic considerations of orbital dynamics. In the case of binaries, we compute the positions of each object at a given time via

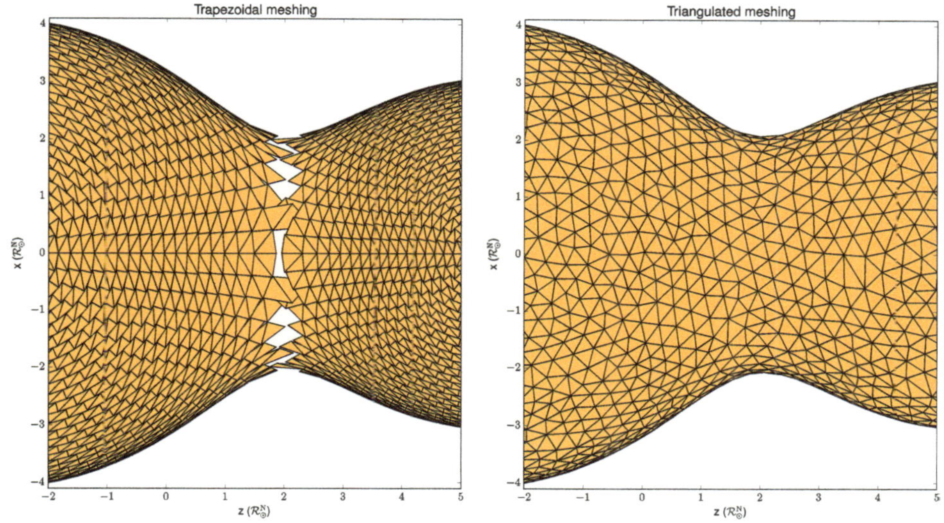

Figure 7.6. Comparison of the neck region coverage in a trapezoidal (left) and triangulated mesh (right) of a contact system with mass ratio $q = 0.5$ and fillout factor $FF = 0.9$. Prša et al. (2016) © 2016. The American Astronomical Society.

Keplerian orbits (Thornton and Marion 2004). The orbits depend on the semimajor axis a, orbital period P_0, mass ratio q, eccentricity e, and systemic velocity v_γ; their orientation is given by an inclination i, argument of periastron ω, and longitude of the ascending node Ω. These parameters describe the dynamical positions of stars in their orbits; however, because of the finite speed of light, these positions do not coincide with where the observer sees the stars.

To address this, we iteratively determine the point in each star's orbit that corresponds to the position of the star at the time the emitted photon crosses the barycenter of the system. Thus, when the stars are closer to the observer, they will appear further along the orbit than their dynamical position; conversely, when the stars are farther from the observer, they will appear retarded along the orbit with respect to their dynamical position.

This change of position due to the Rømer delay allows us to treat the speed of light as infinite and the orbits as dynamical. Adopting the barycenter as reference, note that any radial barycentric motion will result in the anomalous orbital period. Any transverse barycentric motion, on the other hand, will result in the BATV effect, discussed in Section 6.2.1.

7.3 Computation of Local Geometric Quantities

Once the surfaces have been discretized, we need to compute the values of geometrical and radiative parameters at discretized elements. As they depend on position on the stellar surface, they are *local*, defined with respect to a reference point. Any point on the stellar surface could be used as a reference point, yet the most natural choice is the polar radius r_{pole} (see Figure 7.7) because it does not depend significantly on the morphology of the given binary.

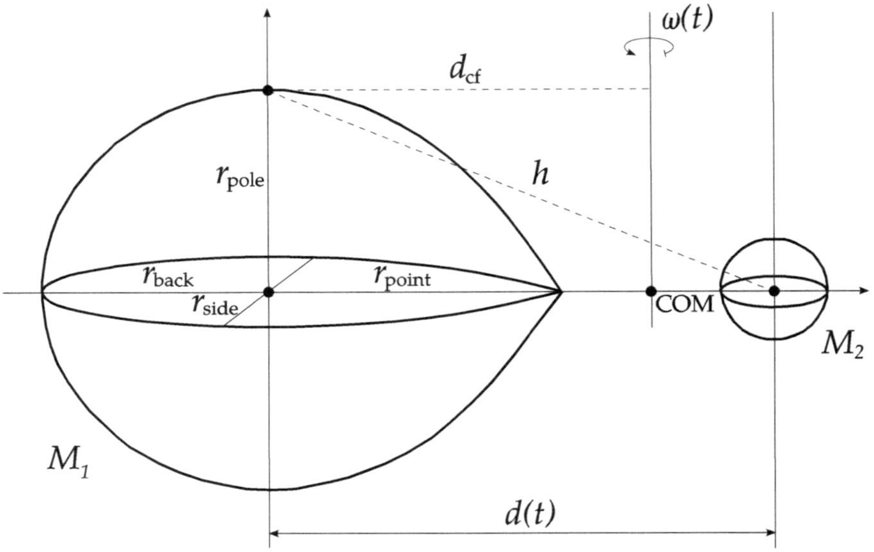

Figure 7.7. Schematic view of a semi-detached binary. There are four reference radii labeled on the lobe-filling binary (left): r_{pole}, r_{back}, r_{side}, and r_{point}; the semi-detached morphology is depicted to amplify the difference between these four radii. When a binary orbit is elliptical, the instantaneous separation $d(t)$ between the two components and the angular velocity $\omega(t)$ are functions of time (phase). This implies that the four radii are phase-dependent as well, bound by the constraint that the volume of each component must by assumption be preserved at every phase.

Local intrinsic quantities, such as local effective temperature or local surface brightness, depend on the local effective gravitational field. Below we outline the procedure that computes all local geometric properties. The results are valid for both circular and elliptical orbits, the only difference being that the first three steps are phase-independent in the case of circular orbits and apply to the periastron phase in the case of elliptical orbits. In the description we will use a primary star with $T_{eff} = 6000$ K, $q = 1$, $F = 1$, and $\Omega = 3.8$ as an example. The critical (L_1) value of the surface potential is 3.75, so this example features a *near*-lobe-filling configuration. Alignment of the axes of rotation and revolution is assumed.

Before we plunge into the nitty-gritty details, we need to make one important distinction. The computation of local quantities in *vertices* is not the same as the computation of local quantities in *surface elements*. Because of the offsetting of the surface elements (discussed in Section 7.1.3), it is the *offset* vertices that span the surface elements, not the original vertices. Yet the local quantities assigned to the surface element are computed by interpolating from the quantities assigned to the original (rather than offset) vertices. This is because the vertices are on the correct equipotential already, and the surface elements are offset from their original position below the equipotential to the new position, where about half is below and about half is above the equipotential. We say about half because the exact partition depends on the underlying curvature of the equipotential. Thus, be mindful of the distinction when we talk about the vertices and when we talk about the surface elements, as those are technically two separate meshes. To make the distinction clearer, we will use a superscript ^{v} preceding a symbol, i.e., $^v\varrho$, to stress that the quantity pertains to the vertex. With that out of the way, let us make the plunge.

STEP 1. COMPUTE REFERENCE VALUES AT PERIASTRON FOR A GIVEN STAR.

By convention, all parameters describe the stars at periastron. To build the model, we need to compute all reference values for each star and use them subsequently to compute the spatial distribution and temporal evolution of all quantities that describe their geometric and thermodynamic properties.

STEP 1A. COMPUTE POLAR RADIUS $\varrho_{\rm pole}$ OF THE GIVEN STAR.

Following the discussion in Section 3, polar radius $\varrho_{\rm pole}$ is computed by solving Equation (3.35) for $\varrho_{\rm pole}$ from the given value of the surface potential Ω. The expression cannot be analytically inverted, so it needs to be solved iteratively.

STEP 1B. COMPUTE THE POLAR GRADIENT MAGNITUDE $|\nabla\Omega_{\rm pole}|$.

Effective gravitational acceleration is one of the principal quantities that determines the distribution of surface brightness over the stellar disks. To compute its reference polar value, we need to evaluate the gradient of the gravitational potential at the star's pole:

$$\Upsilon_{\rm pole} \equiv |\nabla\Omega_{\rm pole}| = \sqrt{\left(\frac{\partial\Omega}{\partial x}\right)^2 + \left(\frac{\partial\Omega}{\partial y}\right)^2 + \left(\frac{\partial\Omega}{\partial z}\right)^2}, \tag{7.9}$$

where partial derivatives are given by Equations (3.41). Evaluating them at the star's pole ($\theta = 0$) yields:

$$\begin{aligned}
\frac{\partial\Omega}{\partial x} &= \frac{q\delta}{(\delta^2 + \varrho_{\rm pole}^2)^{3/2}} - \frac{q}{\delta^2}, \\
\frac{\partial\Omega}{\partial y} &= 0, \\
\frac{\partial\Omega}{\partial z} &= -\frac{1}{\varrho_{\rm pole}^2} - \frac{q\varrho_{\rm pole}}{(\delta^2 + \varrho_{\rm pole}^2)^{3/2}},
\end{aligned} \tag{7.10}$$

where δ is the instantaneous separation between the two stars, normalized to the semimajor axis. The gravitational acceleration $\Upsilon_{\rm pole}$ is unitless, as both $\varrho_{\rm pole}$ and δ are unitless.

STEP 1C. COMPUTE POLAR GRAVITY ACCELERATION $g_{\rm pole}$.

The next step is to compute gravitational acceleration $g_{\rm pole}$ in physical units, which will in turn provide a scaling constant for the associated unitless value $\Upsilon_{\rm pole}$. We do this by noting that, per Equation (3.15),

$$\nabla_r\psi = -\frac{GM_1}{a}\nabla_\varrho\Omega, \tag{7.11}$$

where ∇_r is the gradient operator in absolute units and $\nabla_\varrho \equiv \nabla_r/a$ is the gradient operator in relative units. Thus,

$$\nabla_r \psi = -\frac{GM_1}{a^2} \nabla_r \Omega. \tag{7.12}$$

The acceleration in absolute units, GM_1/a^2, is the scaling factor between $g_{\rm pole}$ and $\Upsilon_{\rm pole}$:

$$\zeta = \frac{g_{\rm pole}}{\Upsilon_{\rm pole}} \equiv \frac{g_{\rm pole}}{|\nabla \Omega_{\rm pole}|}. \tag{7.13}$$

STEP 1D. COMPUTE THE VOLUME AND VOLUME DERIVATIVE.

The volume of an equipotential and its derivative with respect to the potential, $\partial V / \partial \Omega$, are essential because of the volume conservation assumption for eccentric binaries. An efficient way to compute the volume and volume derivative is presented by Prša et al. (2016). The key idea is to write the equipotential Ω in rescaled cylindrical coordinates:

$$\begin{aligned}\tilde{\Omega}(\xi, \sigma, \phi) = \delta\Omega(x, y, z) = \frac{1}{\sqrt{\xi^2 + \sigma}} + \\ + q\left(\frac{1}{\sqrt{\sigma + (\xi - 1)^2}} - \xi\right) + \frac{1}{2}B^2(\xi^2 + \sigma\cos^2\phi),\end{aligned} \tag{7.14}$$

where $\xi = x/\delta$, $\sigma = \varrho/\delta$, and ϕ is the longitude angle; thus, $x = \delta\xi$, $y = \delta\sqrt{\sigma}\cos\phi$, and $z = \delta\sqrt{\sigma}\sin\phi$. $B^2 = F^2(1 + q)\delta^3$ is an auxiliary constant. Volume is then computed as:

$$V = 2\delta^3 \int_{\xi_{\rm back}}^{\xi_{\rm point}} \int_0^{\pi/2} \sigma(\xi, \phi) d\phi d\xi, \tag{7.15}$$

and volume derivative with respect to the potential can be written as:

$$\frac{\partial V}{\partial \tilde{\Omega}} = 2\delta^3 \int_{\xi_{\rm back}}^{\xi_{\rm point}} \int_0^{\pi/2} \left[\frac{\partial \tilde{\Omega}(\xi, \sigma, \phi)}{\partial \sigma}\right]^{-1} d\phi d\xi. \tag{7.16}$$

Integration limits $\xi_{\rm back}$ and $\xi_{\rm point}$ correspond to the rescaled radii $r_{\rm back}$ and $r_{\rm point}$ (see Figure 7.7), and those can be readily computed in the same way as $\varrho_{\rm pole}$ in step 1A. The derivative $\partial\tilde{\Omega}/\partial\sigma$ can be evaluated from Equation (7.14):

$$\frac{\partial \tilde{\Omega}}{\partial \sigma} = -\frac{1}{2}(\xi^2 + \sigma)^{-3/2} - \frac{q}{2}(\sigma + (\xi - 1)^2)^{-3/2} + \frac{1}{2}B^2\cos^2\phi. \tag{7.17}$$

Equations (7.15) and (7.16) are efficiently calculated using an off-the-shelf Legendre–Gauss Quadrature method; see, for example, Širca and Horvat (2012).

STEP 1E. COMPUTE THE SURFACE AREA OF THE STAR.

In much the same vein as in the previous step, the surface of the star could in principle be computed by summing the areas of the discretized surface elements, but this is not optimal. As many aspects of the computation critically depend on the surface area (luminosity, horizon, eclipses, to name just a few), we need to compute it accurately, without depending on the number of surface elements.

Using the same parameterization as in the previous step, surface area is given by:

$$A = 4\delta^2 \int_{\xi_{\rm back}}^{\xi_{\rm point}} \int_0^{\pi/2} \sqrt{\sigma + \frac{1}{4\sigma}\left(\frac{\partial\sigma}{\partial\phi}\right)^2 + \frac{1}{4}\left(\frac{\partial\sigma}{\partial\xi}\right)^2}\, d\phi d\xi, \tag{7.18}$$

where

$$\frac{\partial\sigma}{\partial\xi} \equiv -\frac{\partial\tilde{\Omega}}{\partial\xi}\left(\frac{\partial\tilde{\Omega}}{\partial\sigma}\right)^{-1} \text{ and } \frac{\partial\sigma}{\partial\phi} \equiv -\frac{\partial\tilde{\Omega}}{\partial\phi}\left(\frac{\partial\tilde{\Omega}}{\partial\sigma}\right)^{-1}. \tag{7.19}$$

The integral in Equation (7.18) has the same functional form as volume integrals (Equations (7.15) and (7.16)), and it can be computed using the same methods.

STEP 2. START THE LOOP.

With the reference values computed, the algorithm proceeds to compute local intrinsic quantities at every vertex on the stellar surface. If the orbits are elliptical, all reference quantities computed in step 1 ($\varrho_{\rm pole}$, $\Upsilon_{\rm pole}$, $\partial V/\partial\tilde{\Omega}$, and A) are phase-dependent and need to be recomputed at every phase. The only constraint to which both components by assumption must adhere is the preservation of the volume throughout the period cycle (see Section 3.4.4). In reality, the volume is not strictly preserved, but the errors induced by this approximation are practically negligible (Wilson 1993).

STEP 2A. COMPUTE THE LOCAL RADIUS ${}^{v}\varrho(\theta, \phi)$.

Any vertex on the stellar surface is fully determined by the given surface potential and the chosen direction (θ, ϕ). To compute the local radius, the model solves Equation (3.36) numerically by the Newton–Raphson method. Figure 7.8 depicts the results of such computation for the example star. The point radius is the largest, and the polar radius is the smallest; the two differ by more than 25%.

STEP 2B. COMPUTE THE LOCAL GRADIENT MAGNITUDE ${}^{v}\|\nabla\Omega(\theta, \phi)\|$.

The computation of the local gradient in every vertex yields the local gravitational acceleration ${}^{v}\Upsilon(\theta, \phi)$. The model makes use of the gradient derivatives given by Equation (3.41), where x, y, and z are given by Equation (3.9) and ${}^{v}\varrho(\theta, \phi)$ was obtained in step 2A.

STEP 2C. COMPUTE THE LOCAL GRAVITY ACCELERATION ${}^{v}g(\theta, \phi)$.

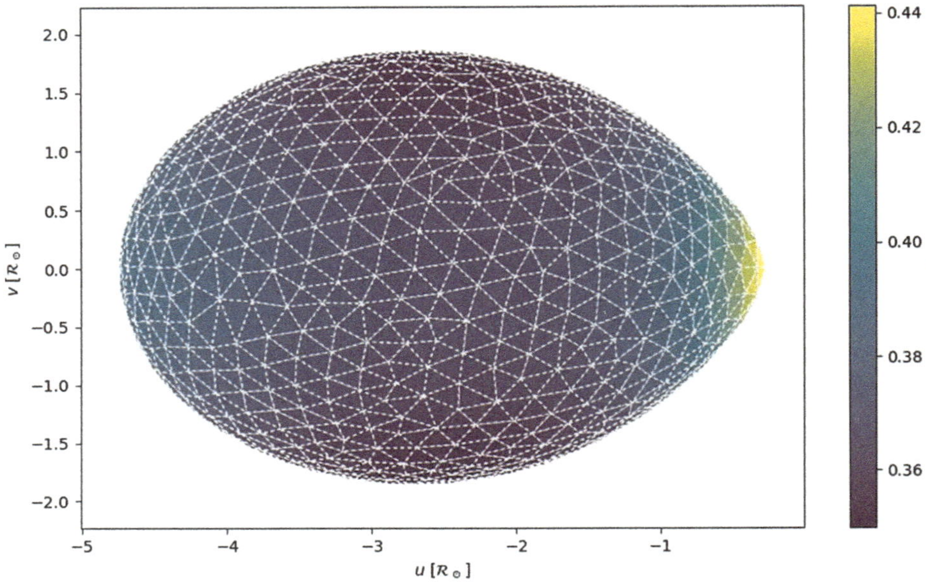

Figure 7.8. Distribution of local radii of the distorted star in a circular orbit. The interactive version of the book features a 360° rotation of the star as it moves through the orbit. The animation is available online at https://doi.org/10.1088/978-0-7503-1287-5.

Instead of repeating the lengthy and time-consuming computation outlined in step 1C, here the algorithm makes use of the scaling factor ζ, given by Equation (7.13):

$$
{}^{v}g(\theta, \phi) = \zeta\ {}^{v}\Upsilon(\theta, \phi) = \frac{{}^{v}\Upsilon(\theta, \phi)}{\Upsilon_{\rm pole}} g_{\rm pole}. \tag{7.20}
$$

It is common to take the logarithm of the obtained value, since gravity acceleration is by convention described in terms of $\log {}^{v}g$ instead of ${}^{v}g$ (see Figure 7.9). The difference in ${}^{v}g$ between the point radius and the pole radius is more than threefold.

STEP 2D. COMPUTE THE GRAVITY DARKENING CORRECTION ${}^{v}\mathcal{G}(\theta, \phi)$.

Gravity darkening describes a distribution of local temperatures across the stellar surface due to tidal and rotational deformation. It is modeled as a power law of effective gravitational acceleration with the slope coefficient β (see Equation (5.4)):

$$
{}^{v}\mathcal{G}(\theta, \phi) = \left(\frac{{}^{v}g(\theta, \phi)}{g_{\rm pole}}\right)^{\beta}. \tag{7.21}
$$

While we compute this correction here, we use it in step 5A to obtain the temperature distribution over the stellar surface.

STEP 2E. COMPUTE THE SURFACE ANGLE ${}^{v}\gamma(\theta, \phi)$.

Because of the tidal and rotational deformation, the normal to the surface and the radius vector at the surface are not aligned. That affects the

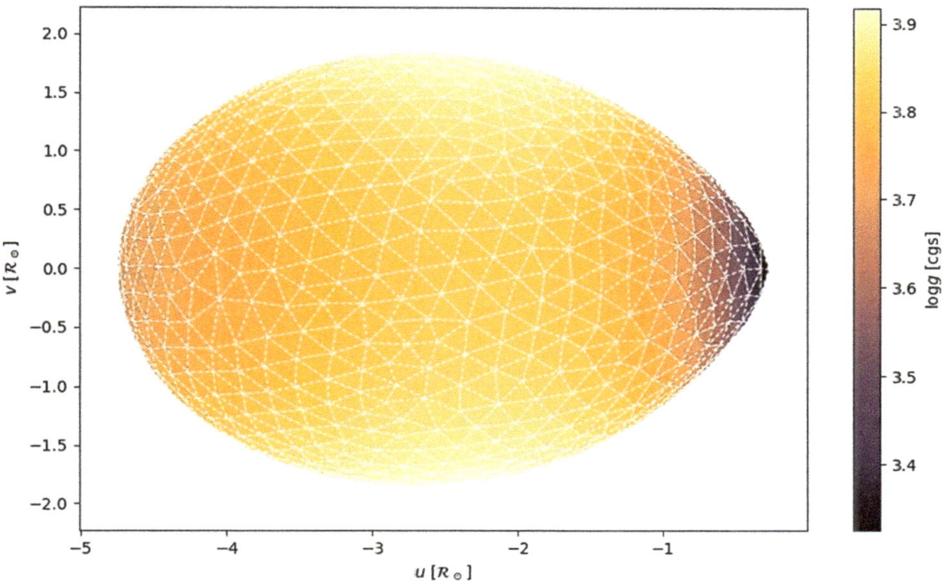

Figure 7.9. Distribution of local effective gravitational acceleration (measured as log g in cgs units) over the surface of the distorted star. The interactive version of the book features a 360° rotation of the star as it moves through the orbit. The animation is available online at https://doi.org/10.1088/978-0-7503-1287-5.

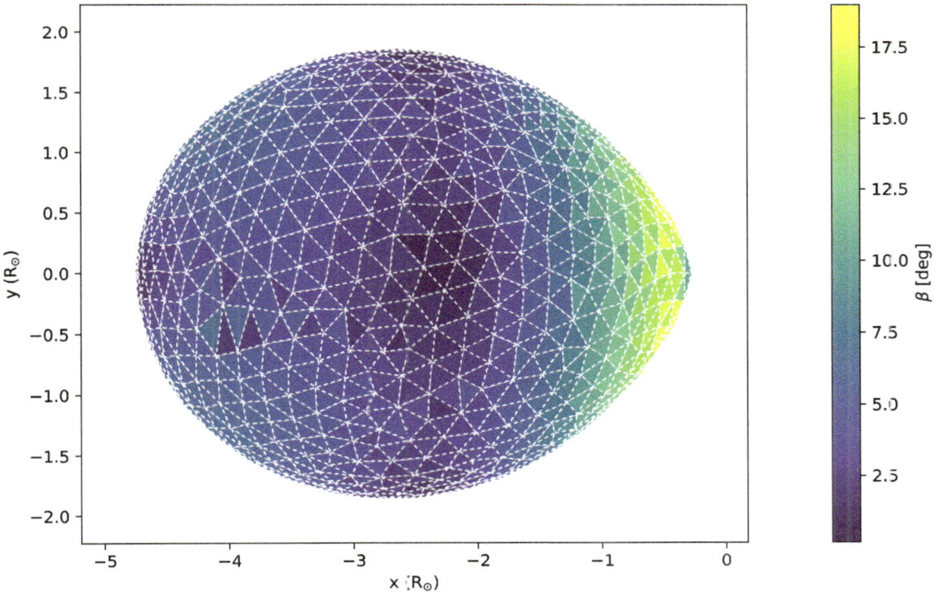

Figure 7.10. Distribution of the angle γ between the surface normal and the radius vector for the lobe-filling star. The jitter is due to the low number of surface elements that are used for depiction purposes.

projection of local surfaces and normal emergent passband intensities toward the observer. The angle ${}^{v}\gamma(\theta, \phi)$ is computed by Equation (3.39):

$$\cos {}^{v}\gamma(\theta, \phi) = {}^{v}\hat{\boldsymbol{n}}(\theta, \phi) \cdot {}^{v}\hat{\boldsymbol{\varrho}}(\theta, \phi). \tag{7.22}$$

Figure 7.10 depicts the values of ${}^{v}\gamma$ for the distorted star. The largest offset close to the point radius approaches 20°.

This concludes the computation of all local geometrical quantities per vertex on the original mesh. The next series of steps is to compute the relevant quantities of the offset mesh.

STEP 3. COMPUTE THE OFFSET MESH.

If we were to compute the surface area by summing the areas of triangles spun by the vertices, it would be smaller than the analytical surface area computed in step 1E because all such triangles would lie below the equipotential. Instead, we offset the vertices along the normal by using Equation (7.8) such that the triangles spun by the offset vertices have areas the sum of which is exactly equal to the analytical surface area of the equipotential. These triangles optimally approximate the true equipotential surface; we refer to them as surface elements.

STEP 3A. COMPUTE THE SURFACE ELEMENT LOCAL RADIUS $\varrho(\theta, \phi)$.

The radius vector to the surface element is computed as the average of the radius vectors to the three offset vertices that span the triangle:

$$\varrho(\theta, \phi) = \frac{1}{3}\|\boldsymbol{\varrho}_1 + \boldsymbol{\varrho}_2 + \boldsymbol{\varrho}_3\|. \tag{7.23}$$

STEP 3B. COMPUTE THE SURFACE ELEMENT NORMAL $\hat{\boldsymbol{n}}(\theta, \phi)$.

The offset normal is computed by a normalized cross product between vectors that span any of the two sides of the triangle:

$$\hat{\boldsymbol{n}}(\theta, \phi) = \frac{\boldsymbol{t}_1 \times \boldsymbol{t}_2}{\|\boldsymbol{t}_1 \times \boldsymbol{t}_2\|}. \tag{7.24}$$

STEP 3C. COMPUTE THE SURFACE ELEMENT ANGLE $\gamma(\theta, \phi)$.

As we offset each vertex according to the underlying curvature along the normal, the angle between the normal and the radius vector will in general differ from ${}^{v}\gamma(\theta, \phi)$ computed in step 2E. We compute the angle the same way as before:

$$\cos \gamma(\theta, \phi) = \hat{\boldsymbol{n}}(\theta, \phi) \cdot \hat{\boldsymbol{\varrho}}(\theta, \phi). \tag{7.25}$$

STEP 3D. COMPUTE LOCAL SURFACE ELEMENT AREAS $\Delta A(\theta, \phi)$.

Once the vertices of all surface elements are computed, we can compute their areas:

$$\Delta A(\theta, \phi) = \frac{1}{2}\|t_1 \times t_2\|, \tag{7.26}$$

where, as before, t_1 and t_2 are the vectors that span two sides of the surface element.

Having computed all local geometrical quantities, the algorithm proceeds with computing local radiative properties in the vertices of each surface.

7.4 Computation of Local Radiative Quantities

Once the local geometry is computed, we can compute the distribution of other intrinsic properties of the modeled objects. Their computation is not in any way distinct from the computation of geometrical quantities; the reason why we present them separately is only to make the overall algorithm logic more transparent. That is why we continue with the enumeration where we left off.

STEP 4. COMPUTE RADIATIVE REFERENCE VALUES AS PERIASTRON.

As was the case for geometric quantities, we compute the reference values for radiative quantities at periastron.

STEP 4A. COMPUTE THE POLAR SURFACE TEMPERATURE.

Binary stars in eccentric orbits change not only their shape as a function of orbital phase Φ but also their surface temperature distribution; it is particularly important to clearly specify what we mean by *effective temperature*: it is the *mean* temperature defined with respect to the bolometric luminosity of the star:

$$L_{\mathrm{bol}} = \sigma \int_{\partial V} T^4(\theta, \phi) dA = \sigma T_{\mathrm{eff}}^4 \int_{\partial V} dA, \tag{7.27}$$

where $T(\theta, \phi)$ is the local surface temperature. The model internally uses the *polar* surface temperature as reference, and the algorithm needs to compute it from the effective temperature by substituting $T^4(\theta, \phi) = {}^{\mathrm{v}}\mathcal{G}(\theta, \phi) T_{\mathrm{pole}}^4$ in Equation (7.27):

$$T_{\mathrm{pole}} = T_{\mathrm{eff}} \left(\frac{\int_{\partial V} dA}{\int_{\partial V} {}^{\mathrm{v}}\mathcal{G}(\theta, \phi) dA} \right)^{1/4} \equiv T_{\mathrm{eff}} \left(\frac{A}{\int_{\partial V} {}^{\mathrm{v}}\mathcal{G}(\theta, \phi) dA} \right)^{1/4}, \tag{7.28}$$

where local gravity-darkening coefficients ${}^{\mathrm{v}}\mathcal{G}(\theta, \phi)$ have been computed in step 2D and surface area A has been computed in step 1E. The discretized form of Equation (7.28) is then:

$$T_{\text{pole}} = \left(\frac{A}{\sum^{v} \mathcal{G}(\theta, \phi) \Delta A(\theta, \phi)} \right)^{1/4} T_{\text{eff}}, \tag{7.29}$$

where $\Delta A(\theta, \phi)$ have been computed in step 3D and the sum goes over all surface elements.

There are two reasons why polar surface temperature is not chosen as the input parameter: (1) there is a notable difference between the observed temperature (i.e., the one that corresponds to the color index) and the polar temperature, so one would need to apply a correction to compare the two; (2) for binaries on eccentric orbits, the polar surface temperature can vary significantly, whereas the effective temperature does not.

STEP 4B. COMPUTE THE POLAR PASSBAND INTENSITY $I_{\text{pb}}^{\text{pole}}$.

Passband intensity has been discussed in detail in Section 4. In short, local normal emergent intensity depends on many factors that describe the hydro- and thermodynamical circumstances in stellar atmospheres, most notable being the local temperature, gravity acceleration, heavy metal abundance, and microturbulent velocity. What we measure on Earth is further impacted by extraneous factors such as Doppler boosting, interstellar extinction, third-light contamination, etc. To account for all these effects, we need to compute an accurate stellar atmosphere model, which is a computationally extensive task, and dedicated codes exist that do just that (e.g., Hubeny and Lanz 1995; Hauschildt et al. 2003; Castelli and Kurucz 2004). These models compute spectral energy distribution (SED) as a function of the aforementioned parameters and as a function of direction, $\mu = \cos\theta$. To obtain (energy-weighted) emergent passband intensity, we need to multiply the SED by the passband response function and integrate it:

$$I_{\text{pb}}(T_{\text{eff}}, \log g, [\text{M/H}],...) = \int_{\lambda} \mathcal{S}(\lambda; T_{\text{eff}}, \log g, [\text{M/H}],...) \mathcal{P}(\lambda) d\lambda. \tag{7.30}$$

The choice of an atmosphere model depends considerably on the type of star and on the intended use. For example, plane-parallel model atmospheres might be perfectly adequate for Sun-like stars, whereas one would likely opt for spherical, non-local thermodynamic equilibrium (NLTE) models for hot O- and B-type stars. Figure 7.11 depicts a comparison between the Castelli and Kurucz (2004), original Kurucz (1993), and blackbody atmospheres for a range of log g and abundance values (denoted by different color tones). The difference between the two model atmospheres at higher effective temperatures is ~12%, and in excess of ~200% at the lowest temperatures (not depicted in the figure to highlight the variation details rather than the overall amplitude). The flatness beyond ~8000 K implies that *relative* flux computations will vary

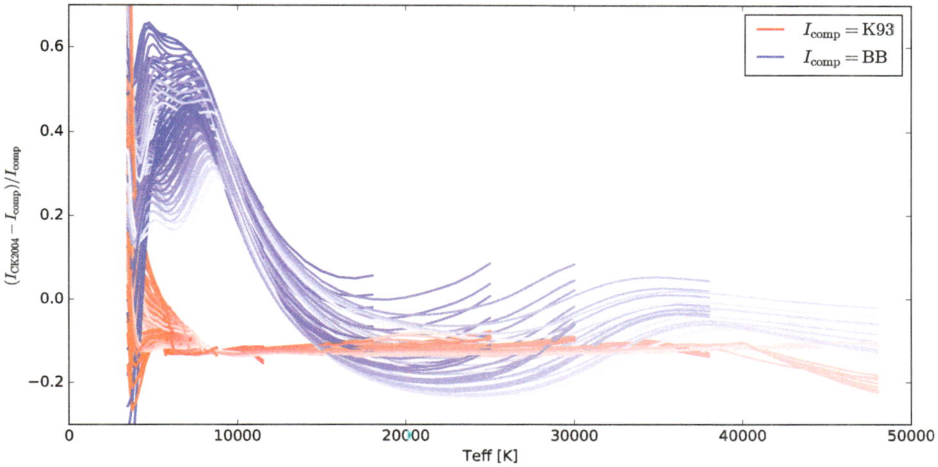

Figure 7.11. Relative difference between the normal emergent passband intensities computed by the Castelli and Kurucz (2004) model atmospheres, the original Kurucz (1993) model atmospheres (red), and blackbody atmospheres (blue). Color tones correspond to the value of $\log g$, where the darkest tone corresponds to the lowest $\log g$ and the lightest tone corresponds to the highest $\log g$ value. Each line corresponds to a combination of $\log g$ and metal abundance. Model atmosphere flux differences are most notable on the low-temperature end (below ~8000 K), in excess of 200% for M stars. Blackbody fluxes deviate across the entire temperature range. Taken from Prša et al. (2016). © 2016. The American Astronomical Society.

to the extent of the deviation from the flat curve, but the *absolute* flux computations will differ by as much as ~15% (Prša et al. 2016).
To save on computing time, normal emergent intensities for individual passbands can be tabulated ahead of time and stored for future lookup use. These tables span all atmospheric parameters that influence the computed SED functions. This allows the model to skip the integration of a SED function (be it blackbody or a model atmosphere) and reduce the required operations to an nD interpolation.

STEP 5. START THE LOOP.

STEP 5A. COMPUTE LOCAL TEMPERATURE ${}^{v}T(\theta, \phi)$.

In this step we distribute the local temperatures according to the gravity-darkening law:

$$ {}^{v}T(\theta, \phi) = {}^{v}\mathcal{G}^{1/4}(\theta, \phi) T_{\text{pole}}. \tag{7.31} $$

Figures 7.12 and 7.13 depict a local temperature distribution across the surface of a radiative star ($\beta = 1.0$) and of a convective star ($\beta = 0.32$), respectively. Gravity darkening causes equatorial regions to be cooler than polar regions. As expected, the effect is more pronounced for radiative envelopes.

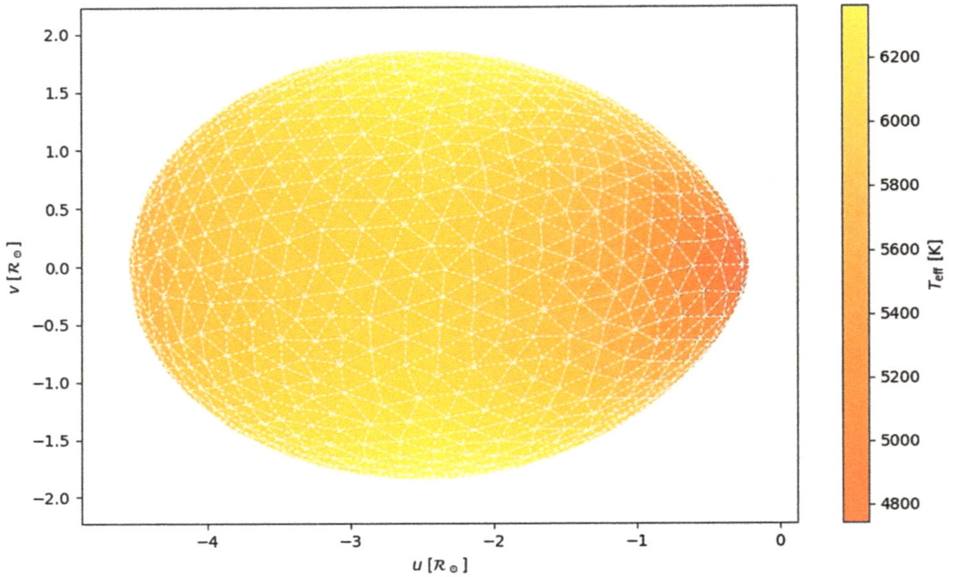

Figure 7.12. Distribution of local temperature over the surface of the distorted star assuming a radiative envelope. An animation is available online at https://doi.org/10.1088/978-0-7503-1287-5.

Note that we should not apply any other corrections at this point (such as spots, reflection, etc.) because of the definition of passband luminosity. We need an *intrinsic* measure of the star's luminosity before we apply further corrections. Otherwise, the luminosity would become aspect dependent, which is certainly not beneficial.

STEP 5B. COMPUTE LOCAL NORMAL PASSBAND INTENSITY ${}^{\nu}I_{\rm pb}(\theta, \phi)$.

Given the distribution of local temperatures computed in step 5A and local effective gravities $\log g$ computed in step 2C, computing ${}^{\nu}I_{\rm pb}(\theta, \phi)$ boils down to querying the atmosphere model lookup tables.

STEP 5C. COMPUTE LOCAL RADIAL VELOCITIES $rv(\theta, \phi)$.

We can compute RVs in two regimes: dynamical and integrated. Dynamical RVs correspond to the line-of-sight component of the orbital velocity vector of each star; as they are analytically computed, they do not exhibit any proximity effects (i.e., the Rossiter–McLaughlin effect; see Section 3.4.7). In order to model these effects, RVs must also be sampled across the mesh of each star. Local RVs are assigned to each element by computing the (gravitationally shifted) projected velocity per Equations (3.120) and (3.121).

STEP 6. COMPUTE THE SCALE BETWEEN INTENSITY AND LUMINOSITY.

The passband intensity distribution computed in step 5B is given in internal units. These internal units now need to be scaled to agree with the requested

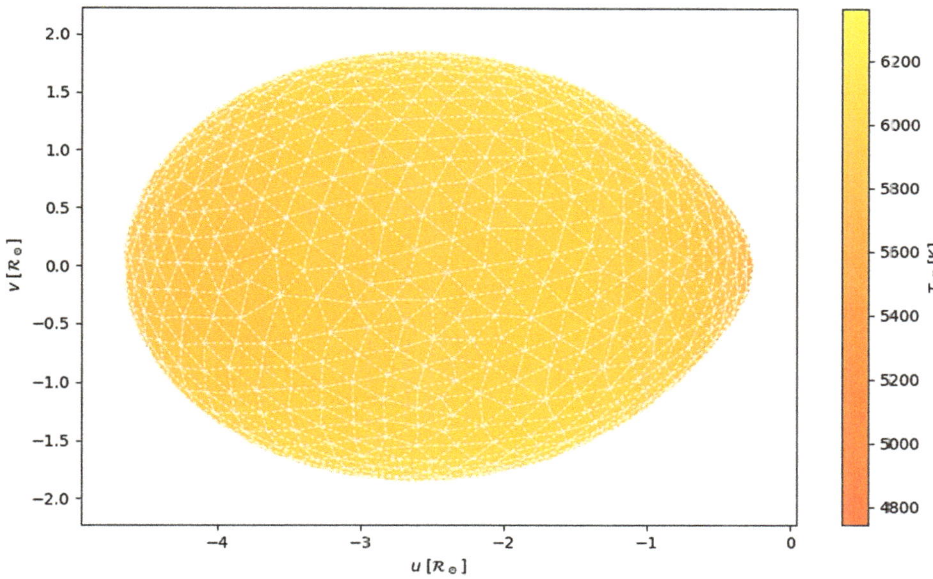

Figure 7.13. Distribution of local temperature over the surface of the distorted star assuming a convective envelope. An animation is available online at https://doi.org/10.1088/978-0-7503-1287-5.

passband luminosity units. To do this, the model evaluates Equation (4.29) to obtain passband luminosity in internal units. The scale is then obtained by:

$$\zeta_I \equiv \frac{L_{\mathrm{pb}}}{L_{\mathrm{pb}}^{\mathrm{internal}}}, \tag{7.32}$$

and all intensities are multiplied by this scale. Evaluating Equation (4.29) will then yield the correct passband luminosity, and all subsequent operations on intensities (i.e., spots, reflection, etc.) will retain the correct scaling.

STEP 7. COMPUTE EXTRINSIC EFFECTS.

With the scale set, we can now account for any other effects that change the temperature and emergent intensity distribution across the stellar surfaces.

STEP 7A. ADD SPOTS TO THE STELLAR SURFACE.

Any other effects that modify the temperature distribution across the surface, such as spots, are accounted for here. The local temperature of the vertex covered by a spot is determined by the spot's temperature factor:

$$^{v}T_s(\theta, \phi) = \tau_{\mathrm{sp}}\, {}^{v}T(\theta, \phi), \tag{7.33}$$

where ${}^{v}T(\theta, \phi)$ is the underlying temperature of the vertex.

STEP 7B. COMPUTE THE REFLECTION EFFECT CORRECTION $\mathcal{R}(\theta, \phi)$.

The main difference between the reflection effect and other surface brightness phenomena computed in the previous loop is that a given surface element on each star receives integrated radiation from all visible surface elements on the other star. That is why the distribution of surface temperature and intensity need to be computed before reflection is treated. The computational cost for the accurate treatment of reflection scales with the square of the number of surface elements, so computing it for dense meshes can become the dominant computational expense of the model.

For circular, synchronous orbits, reflection does not depend on the orbital phase: it only needs to be computed once per surface element.

To compute reflection, we first need to compute the bolometric limb-darkening correction $\mathcal{L}^{\mathrm{bol}}(\theta, \phi)$. Heating of one stellar surface by another is a bolometric process, i.e., it does not depend on the wavelength of the incident light. To compute the amount of radiation received from the component, we thus need to compute a *bolometric* limb-darkening correction. This correction depends on the angle at which each surface element on one star "sees" the surface elements on the other star. Bolometric limb-darkening coefficients are either passed as parameters or interpolated from limb-darkening tables.

We can then compute local reflection coefficients $\mathcal{R}_1(\theta, \phi)$ and $\mathcal{R}_2(\theta, \phi)$. This is an iterative process that estimates the total bolometric flux $F_{2\to1}^{\mathrm{bol}}(\theta, \phi)$ received from the companion star at the local surface element. We set the reflection coefficient $\mathcal{R}_2$ to its initial value (unity) and compute irradiance per Equation (5.9). Then we reverse the roles of stars and compute irradiance using an initial value of $\mathcal{R}_1$. The updated values of $\mathcal{R}_1$ and $\mathcal{R}_2$ are then computed by Equations (5.12) and (5.13). We use these for the next round of computations in Equation (5.9) until we reach the desired accuracy.

With the scheme outlined above, we have computed all intrinsic locally defined geometrical and radiative quantities of an eclipsing binary system. The total flux intercepted by the observer will be obtained by integrating the flux from each visible surface element in the direction of the observer.

7.5 Computation of Aspect-dependent Quantities

STEP 8. COMPUTE THE HORIZON.

The analytical computation of the horizon has been discussed in Section 3.3.2. Here we need to compute a horizon of a discretized mesh, but Equation (3.86) can be readily used to test the validity of the numerical approach.

Horizon computation is done on the offset mesh because it corresponds to the exact surface area of the surface equipotential. Figure 7.14 depicts the projected limb of the triangulated mesh compared to the theoretical horizon (see

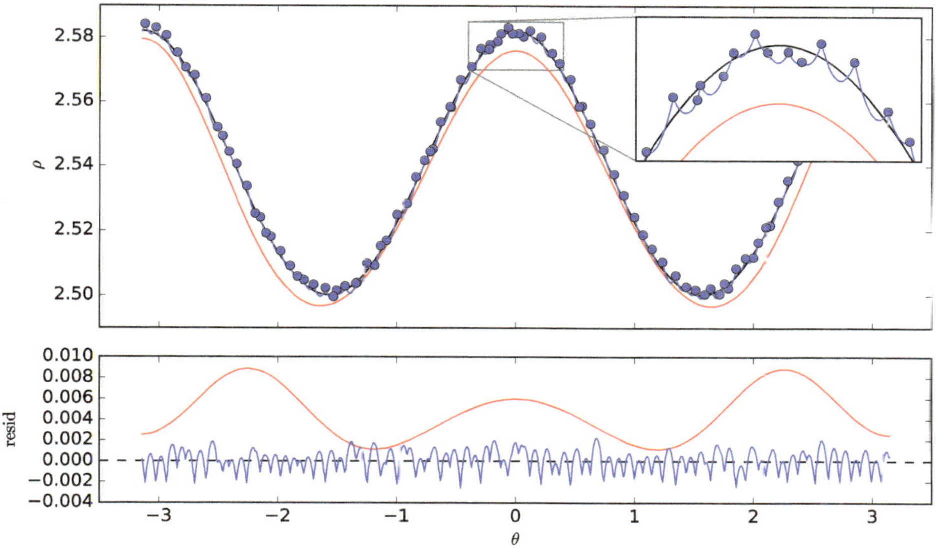

Figure 7.14. Horizon of a Roche star in polar coordinates. The black curve represents the analytic horizon (described in Section 3.3.2), the red curve represents the Fourier fit to the horizon, the blue circles represent surface elements of the offset mesh that are on the horizon, and the blue line represents the polygonal outline that these points span. The bottom panel shows the residuals between the Fourier-based (red) and offset triangulation-based (blue) horizons with respect to the analytical horizon. Adopted from Prša et al. (2016). © 2016. The American Astronomical Society.

Equation (3.86)) and demonstrates that the determined horizon of each star agrees well with the theoretical expectation. In contrast, the Fourier-based horizon method implemented in older codes determines the horizon of each star by fitting a Fourier series to the coordinates of surface elements closest to the horizon (Wilson 1993). In consequence, this approach will always underestimate the size of the limb with respect to the analytic horizon. This underestimation converges with an increase in the number of surface elements, but very slowly. Due to the offsetting of surface elements to obtain the correct numerical surface area, the model does not underestimate the horizon even for a very coarse mesh, and increasing the number of elements only makes the horizon smoother.

STEP 9. COMPUTE THE ECLIPSES.

Eclipse detection of triangular meshes can be performed algebraically (Prša et al. 2016). First, we do back-face culling of triangles, based on the normals of the underlying smooth surface, to obtain a list of potentially visible triangles. Then, we utilize Painter's algorithm (Hughes et al. 2013) on the surface elements that are ordered with respect to distance from the observer and projected onto the plane of sky. Both steps are standard techniques for hidden surface removal in computer graphics. Next, we calculate the visible part of the projected triangles using the Vatti's clipping method (Vatti 1992). From the visible parts of the

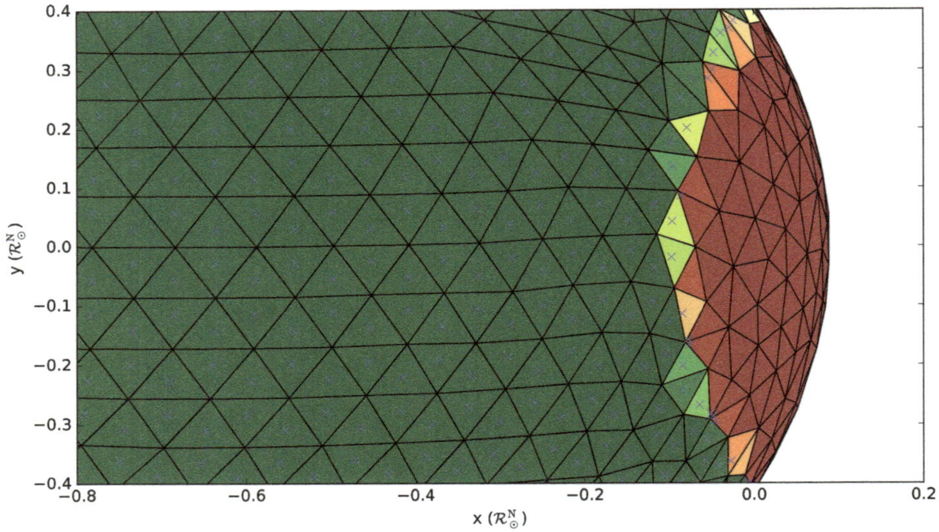

Figure 7.15. Segment of the partially eclipsed mesh. The colors of each triangle represent the ratio of the triangle area that is visible, with green being fully visible and red being fully eclipsed. The blue crosses represent the location of the centroid of the visible portion of the triangle. For visible (green) triangles, the centroids appear at the geometric centers of the triangles, whereas the partially visible (orange and yellow) triangles show the centroid moving away from the eclipsed region. These centroids are used to integrate over the visible surface of the stars. Adopted from Prša et al. (2016). © 2016. The American Astronomical Society.

projected triangles we deduce the ratio of each triangle that is visible, as well as the revised centroid of the visible portion.

Any given *observable* quantity of each triangle is then determined by the weighted mean of that quantity at each vertex to determine the value at the centroid, under the assumption that these properties vary linearly across the local surface element (see Figure 7.15). Its observable area is then determined based on the calculated ratio. This method allows an accurate approximation of ingress/egress without the need for a fine mesh that is computationally much more expensive.

STEP 10. COMPUTE THE OBSERVABLES.

We now have everything prepared to do the final summation of emergent passband intensity in the direction of the observer from all visible elements, and we apply the remaining extrinsic effects such as interstellar extinction and third light.

STEP 10A. COMPUTE LIMB-DARKENING CORRECTION.

The line-of-sight intensity in general differs from the normal emergent intensity because of limb darkening, covered in detail in Section 4.4. To compute the correction, we need to select a limb-darkening model (see Table 4.1) and apply it to all surface elements. Figure 7.16 shows the

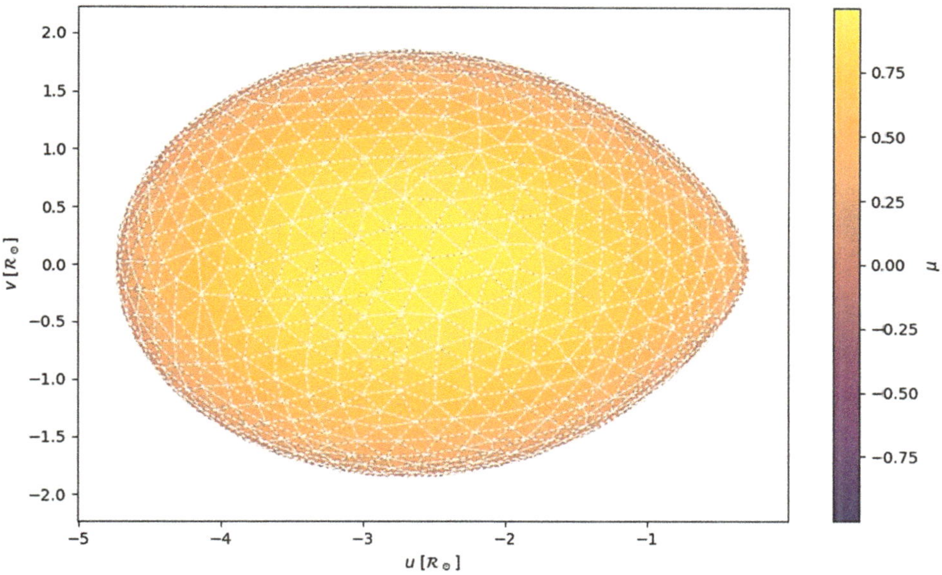

Figure 7.16. Angle φ between the surface normal $\hat{\boldsymbol{n}}$ and the line-of-sight vector $\hat{\boldsymbol{u}}$. If $-90° < \varphi < 90°$ or, alternatively, if $\cos\varphi > 0$, the corresponding surface element will be visible; otherwise, it will not. This condition thus determines the horizon of the star in question. An animation is available online at https://doi.org/10.1088/978-0-7503-1287-5.

distribution of φ, the angle between the surface normal and the line-of-sight vector, that demonstrates the effect of limb darkening on intensity.

STEP 10B. COMPUTE LIGHT CURVES.

Flux is computed by integrating scaled, limb-darkened, projected intensities along the line of sight over the visible surface of the star. Any extraneous light (passed as a parameter) is then added to this value to obtain the observable flux. The process is then repeated for each time stamp, where local quantities are updated if the mesh has changed, and reintegrated to obtain a new flux.

To evaluate the precision of flux computation, we consider the Sun–Earth system. This is a convenient test case because it is the best-calibrated benchmark we have, and because it pushes model parameters to the extreme operation mode. We limit our test to the blackbody model because it is the only model that we can integrate analytically, and we need to do that for comparison purposes. Figure 7.17 summarizes the results. The black line is the theoretical baseline computed by analytical integration. Other curves represent the results from meshed models, and they depict fluxes as a function of the number of surface elements. Offset triangulation performs most robustly, and it reproduces the analytical value effectively.

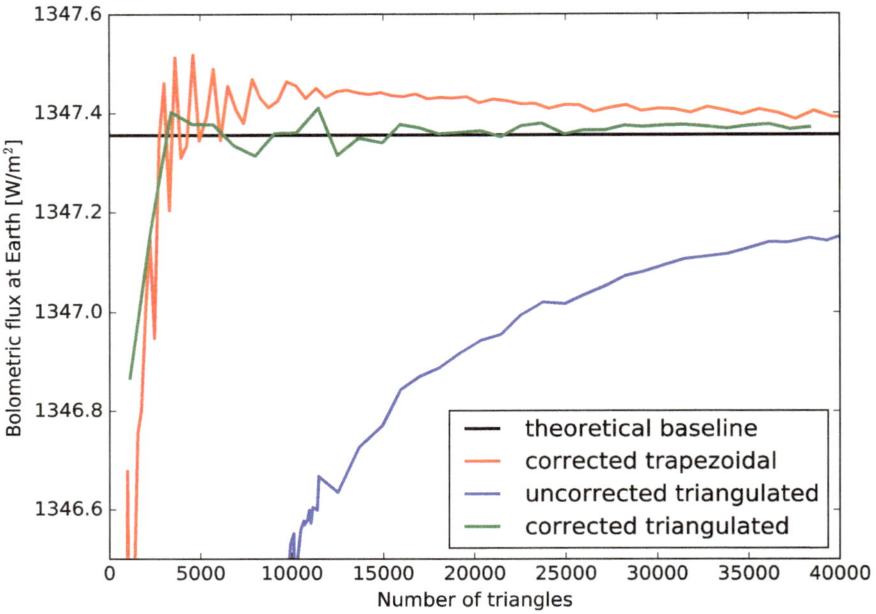

Figure 7.17. Comparison of the fluxes computed analytically (black line) and for different meshes: trapezoids with analytical areas (red line), triangulation (blue line), and offset triangulation (green line) as a function of mesh size. Offset triangulated meshes are consistent with the benchmark and reach it for a smaller number of surface elements than any other method. Prša et al. (2016) © 2016. The American Astronomical Society.

STEP 10C. COMPUTE RADIAL VELOCITY CURVES.

Local RVs computed in step 5C are now assigned to each surface element, weighted by the visible centroid and then integrated over the visible area to the photometrically weighted, integrated RV curve.

STEP 10D. APPLY FINITE INTEGRATION TIME CORRECTION.

When modeling short-period systems, a finite exposure time starts to become relevant. This is especially true for *Kepler* data, where CCDs are read out every 6.54 s (6.02 s live time) and then summed into 29.4244-minute Long Cadence bins. This implies that every exposure has an effective ~30-minute integration time during which any flux variation will be smeared into a single data point. This bears little significance for objects with periods longer than ~10 days, but it becomes important for short-period systems because it convolves the true signal with the exposure time boxcar, resulting in phase smearing. The effect is more pronounced for semi-detached and contact binaries, with orbital periods as short as a few hours. To account for the smoothing of observations due to finite exposure times, we can oversample the light curve and provide an averaged value for any given time.

STEP 10E. APPLY INTERSTELLAR EXTINCTION.

To conclude the computation, we synthesize the composite SED function by co-adding spectral intensities over the visible surfaces, weighted by the element centroids. To this spectrum we apply interstellar extinction laws (see Section 5.5), integrate over the passband wavelength interval, and correct the flux by the computed extinction.

The steps outlined above are only one possible approach to computing the observables from radiating sources. They come with some inherent limitations, the most important of which is the dependence of the integrated flux on the number of surface elements. Ideally, the computed flux should not depend on the computational aspects, such as the fineness of the mesh, but this is an inherent problem for all numerical schemes. Flux values converge to the analytical value linearly ($\propto 1/N$, where N is the number of surface elements), which is quite slow and is further impacted by other second-order effects, most notably limb darkening. The exact convergence properties are difficult to assess because of the nonlinearity of the parameter space, but tests show that for a moderate number of triangles ($\sim$5000) the attained relative accuracy is of the order of 10^{-3}. The impact on *relative* light-curve shapes (i.e., flux ratios) is much smaller because all phase points are affected equivalently, but whenever flux values are sought, their dependence on the mesh size should be kept in mind.

We should also stress that the computational time cost is a nontrivial consideration. While the number of computations per second is ever increasing with the development of modern computing infrastructure, there is still a trade-off between accuracy and computational time cost that needs to be taken into account. The computing-intensive parts should be written in a fast computer language such as C, while the interface should be written in an interpreted language such as Python. That way we can be rest assured that the dominant source of slowdown is the increased model fidelity. Hence, it is likely that the preferred mode of deployment of future eclipsing binary models is on high-performance computing clusters.

References

Castelli, F., & Kurucz, R. L. 2004, arXiv:astro-ph/0405087

Hartmann, E. 1998, VComp, 14, 95

Hauschildt, P. H., Allard, F., Baron, E., Aufdenberg, J., & Schweitzer, A. 2003, in ASP Conf. Ser. 298, GAIA Spectroscopy: Science and Technology, ed. U. Munari (San Francisco, CA: ASP), 179

Hubeny, I., & Lanz, T. 1995, ApJ, 439, 875

Hughes, J. F., van Dam, A., McGuire, M., et al. 2013, Computer Graphics: Principles and Practice (3rd ed.; Boston, MA: Addison-Wesley)

Kallrath, J., & Milone, E. F. 2009, Eclipsing Binary Stars: Modeling and Analysis (New York: Springer)

Kallrath, J., Milone, E. F., Terrell, D., & Young, A. T. 1998, ApJ, 508, 308

Kurucz, R. L. 1993, in ASP Conf. Ser. 44, IAU Colloq. 138, Peculiar versus Normal Phenomena in A-type and Related Stars, ed. M. M. Dworetsky, F. Castelli, & R. Faraggiana (San Francisco, CA: ASP), 87

Milone, E. F., Stagg, C. R., & Kurucz, R. L. 1992, ApJS, 79, 123

Prša, A., Conroy, K. E., Horvat, M., et al. 2016, ApJS, 227, 29

Prša, A., & Zwitter, T. 2005, ApJ, 628, 426
Širca, S., & Horvat, M. 2012, in Graduate Texts in Physics, Computational Methods for Physicists: Compendium for Students (Berlin: Springer)
Taubin, G. 1995, in Proc. Fifth Int. Conf. on Computer Vision (Washington, DC: IEEE Computer Society Press), 902
Thornton, S. T., & Marion, J. B. 2004, Classical Dynamics of Particles and Systems (Pacific Grove, CA: Brooks/Cole)
van Hamme, W., & Wilson, R. E. 2003, in ASP Conf. Ser. 298, GAIA Spectroscopy: Science and Technology, ed. U. Munari (San Francisco, CA: ASP), 323
Van Hamme, W., & Wilson, R. E. 2007, ApJ, 661, 1129
Vatti, B. R. 1992, Commun. ACM, 35, 56
Wilson, R. E. 1979, ApJ, 234, 1054
Wilson, R. E. 1990, ApJ, 356, 613
Wilson, R. E. 1993, in ASP Conf. Ser. 38, New Frontiers in Binary Star Research, Vol. 38, ed. K.-C. Leung, & I.-S. Nha (San Francisco,CA: ASP), 91
Wilson, R. E. 2008, ApJ, 672, 575
Wilson, R. E., & Biermann, P. 1976, A&A, 48, 349
Wilson, R. E., & Devinney, E. J. 1971, ApJ, 166, 605
Wilson, R. E., & Sofia, S. 1976, ApJ, 203, 182
Wilson, R. E., & Van Hamme, W. 2014, ApJ, 780, 151

Chapter 8

Approaches to Solving the Inverse Problem

In the previous chapters we have established that the number of geometrical and physical parameters entering the model is substantial: the complete list includes more than 100 parameters. The task of the model is to take these parameters and compute synthetic curves (light curves, radial velocity curves, spectra, etc.). This is referred to as the *direct* or *forward* problem: given the set of parameters, synthetic curves are *uniquely* determined.

In practice, however, we need to do the exact opposite: given the observed data, we need to determine a set of parameters for which the model yields a matching synthetic curve. This is referred to as the *inverse* problem. In contrast to the direct problem, due to parameter correlation and data noise, there may be many combinations of parameters that yield an indistinguishable synthetic curve within a given accuracy; thus, the inverse problem may *not* have a unique solution. This is referred to as *solution degeneracy*: the "right" combination of the "wrong" parameters may fit the observed data exceptionally well (Coughlin 2010).

The most useful data for modeling purposes are those that come with robust error estimates. When solving the inverse problem, most algorithms assume noncorrelated white noise, which is frequently (read: always) not the case. The uncertainties are typically temporally and spatially correlated, depend on the instrumentation, change with observing conditions, etc. Thus, for a robust model it is not sufficient to properly model the signal; we also need to properly model the noise. Failure to do so will result in incorrect estimates of model parameters—the errors may be frequently underestimated, and parameter values may suffer from systematics.

The effects of degeneracy can be reduced (and in some cases even diminished) by acquiring diverse and accurate observations, using heuristic numerical methods, and, most importantly, employing an ample amount of common sense. In the best cases the accuracy in fundamental parameters (masses and radii) of the obtained solution dips below ~2%–3%. In this chapter we introduce several numerical methods for solving the inverse problem and point out some of the caveats associated with them.

doi:10.1088/978-0-7503-1287-5ch8

8.1 Local and Global Minima in the Parameter Hyperspace

The task of any minimization algorithm employed for solving the inverse problem is to minimize the residuals between observed data and synthetic curves. The residuals are typically quantified in terms of the χ^2 cost function:

$$\chi_k^2 = \sum_{i=1}^{N_k} w_k w_i (F_i^{\rm obs} - F_i^{\rm syn})^2 = \frac{1}{\sigma_k^2} \sum_{i=1}^{N_k} w_i (F_i^{\rm obs} - F_i^{\rm syn})^2 \tag{8.1}$$

for the kth passband, where N_k and σ_k are the number of data points and per-passband standard deviation, respectively, w_i are individual data point weights, $w_k = 1/\sigma_k^2$ are passband weights, $F^{\rm obs}$ is the observed flux at a given phase, and $F^{\rm syn}$ is the computed flux at the same phase. As the weighted variance is given by:

$$s_k^2 = \frac{1}{N_k - 1} \sum_{i=1}^{N_k} w_i (F_i^{\rm obs} - F_i^{\rm syn})^2, \tag{8.2}$$

the passband cost function χ_k^2 may be readily expressed as:

$$\chi_k^2 = (N_k - 1)\frac{s_k^2}{\sigma_k^2}. \tag{8.3}$$

The overall cost function is a sum of individual cost functions for each observed curve:

$$\chi^2 = \sum_{k=1}^{M} \chi_k^2 = \sum_{k=1}^{M} (N_k - 1)\left(\frac{s_k}{\sigma_k}\right)^2, \tag{8.4}$$

where M is the number of curves. If σ_k are properly estimated, the ratio s_k/σ_k is of the order unity and χ^2 of the order $N_{\rm tot} = \sum_k N_k$. We use this *reduced* χ^2 cost function:

$$\chi_\nu^2 := (\chi^2/N_{\rm tot}). \tag{8.5}$$

The χ_ν^2 is a direct measure of the goodness of fit: if the value is ~1, the discrepancy between the model and observations is small and the fit is likely reasonable. If χ_ν^2 is large, the discrepancy is large and either the fit is inadequate or the uncertainties are poorly estimated. Note that we commonly use the word *fit* instead of *solution*, as a good fit to the data does not necessarily imply a good solution.

As we have seen in previous chapters, the model for eclipsing binaries is highly nonlinear, with some strong parameter correlations. Inherent observed data noise adds to this, making the inverse problem difficult to solve. Parameters span a high-dimensional hyperspace that, due to constraints and correlations, has a very complicated topology. This in turn implies that such a hyperspace is expected to have many minima with respect to the reduced cost function χ_ν^2. The discernible property of these minima is their depth: there are many shallow minima that occupy certain areas of the hyperspace that are called *local*: a small displacement in parameter values may knock the fit out of the minimum and enable the minimizer to further reduce the value of the cost function. On the other hand, there is an area of the deepest minimum within the hyperspace that is called *global*: any displacement in parameter values will not result in further improvement of the fit.

Parameter χ_ν^2 may thus be regarded as a qualifier for the depth of the minimum. If a minimization method converges and χ_ν^2 is still large, chances are that the method had converged to a local minimum, where it got stuck. If χ_ν^2 is small, the method got close to the depth of the global minimum. The smaller the χ_ν^2, the deeper the minimum. This is an important point that we will further build on at the end of this chapter, when we discuss how to handle solution degeneracy.

The minimizer's task to objectively discern these minima and to find the deepest one is very demanding: any naive approaches of fitting the whole spectrum of parameters simultaneously will almost surely fail. It is of utmost importance to have a system of control imposed on a minimizer that guides it in its solution-seeking process. We proceed with overviewing the most successful minimization algorithms that have been deployed for solving the inverse problem of eclipsing binaries.

8.2 Differential Corrections

One of the most straightforward numerical methods for driving the χ^2 minimization is the method of differential corrections (DC), first proposed by Euler (1755). The method is based on replacing partial derivatives of the residuals, $f_i \equiv F_i^{\mathrm{obs}} - F_i^{\mathrm{syn}}$, with finite differences:

$$\frac{\partial f_i(\cdots p_k \cdots)}{\partial p_k} \to \frac{f_i(\cdots p_k + \Delta p_k \cdots) - f_i(\cdots p_k \cdots)}{\Delta p_k}, \tag{8.6}$$

where $f_i(\cdots p_k \cdots)$ is the n-dimensional scalar function (the residuals) that we are minimizing. At the expense of calculation time, the accuracy of Equation (8.6) may be slightly improved by using symmetric differences:

$$\frac{\partial f_i(\cdots p_k \cdots)}{\partial p_k} \to \frac{f_i(\cdots p_k + \Delta p_k/2 \cdots) - f_i(\cdots p_k - \Delta p_k/2 \cdots)}{\Delta p_k}. \tag{8.7}$$

Evaluation of this derivative informs the method of the slope along parameter p_k in the parameter hyperspace. Thus, to minimize the residuals, we need to take a suitable step along the slope. The step size should be commensurate with the slope: if the slope is small, the change should be small; if the slope is large, the change is large. The change in the residuals $\xi_i \equiv \Delta(F_i^{\mathrm{obs}} - F_i^{\mathrm{syn}})$ can be written as:

$$\xi_i \approx \sum_k \frac{\partial f_i(\cdots p_k \cdots)}{\partial p_k} \Delta p_k \tag{8.8}$$

for $i = 1 \ldots N$ data points and $k = 1 \ldots K$ adjusted parameters. In all practical situations $N \gg K$, so this is an overdetermined system of N equations with K

unknowns. We solve it by employing the least-squares (Gauss–Newton) method; for notation conciseness, we form the following vectors and matrices:

$$\boldsymbol{\xi} = \begin{pmatrix} \xi_1 \\ \vdots \\ \xi_N \end{pmatrix}, \quad \boldsymbol{\Delta} = \begin{pmatrix} \Delta p_1 \\ \vdots \\ \Delta p_K \end{pmatrix}, \quad \boldsymbol{A} = \begin{pmatrix} \dfrac{\partial f_1}{\partial p_1} & \dots & \dfrac{\partial f_1}{\partial p_K} \\ \vdots & \ddots & \vdots \\ \dfrac{\partial f_N}{\partial p_1} & \dots & \dfrac{\partial f_N}{\partial p_K} \end{pmatrix}. \tag{8.9}$$

The set of Equations (8.8) can then be written concisely as $\boldsymbol{\xi} = \boldsymbol{A}\boldsymbol{\Delta}$. The cost function we are minimizing is:

$$S(\boldsymbol{\Delta}) = \| \boldsymbol{W}^{1/2}(\boldsymbol{\xi} - \boldsymbol{A}\boldsymbol{\Delta}) \|^2, \qquad \boldsymbol{W} = \begin{pmatrix} 1/\sigma_1^2 & & 0 \\ & \ddots & \\ 0 & & 1/\sigma_N^2 \end{pmatrix}, \tag{8.10}$$

where we introduced the matrix $\boldsymbol{W}$ of individual weights $w_i \equiv 1/\sigma_i^2$. The cost function can be expanded as:

$$S(\boldsymbol{\Delta}) = (\boldsymbol{\xi} - \boldsymbol{A}\boldsymbol{\Delta})^T \boldsymbol{W} (\boldsymbol{\xi} - \boldsymbol{A}\boldsymbol{\Delta}) = \boldsymbol{\xi}^T \boldsymbol{W} \boldsymbol{\xi} - 2\boldsymbol{\Delta}^T \boldsymbol{A}^T \boldsymbol{W} \boldsymbol{\xi} + \boldsymbol{\Delta}^T \boldsymbol{A}^T \boldsymbol{W} \boldsymbol{A} \boldsymbol{\Delta}. \tag{8.11}$$

To minimize it, we evaluate a derivative with respect to $\boldsymbol{\Delta}$:

$$\frac{\partial S(\boldsymbol{\Delta})}{\partial \boldsymbol{\Delta}} = -2\boldsymbol{A}^T \boldsymbol{W} \boldsymbol{\xi} + 2\boldsymbol{A}^T \boldsymbol{W} \boldsymbol{A} \boldsymbol{\Delta} = 0. \tag{8.12}$$

That gives us the expression for $\boldsymbol{\Delta}$:

$$\boldsymbol{\Delta} = (\boldsymbol{A}^T \boldsymbol{W} \boldsymbol{A})^{-1} \boldsymbol{A}^T \boldsymbol{W} \boldsymbol{\xi}. \tag{8.13}$$

This might look scary, but it is really not. The components of matrix $\boldsymbol{A}$ (partial derivatives) can be computed either analytically or numerically (by replacing partial derivatives with finite differences per Equation (8.7)), and the components of vector $\boldsymbol{\xi}$ and matrix $\boldsymbol{W}$ are provided by observations, so solving Equation (8.13) boils down to finding the inverse of the matrix $\boldsymbol{A}^T \boldsymbol{W} \boldsymbol{A}$. Books have been written on finding the inverse of large matrices, and any off-the-shelf algorithm that does that is likely to be sufficient for our purposes.

The procedure outlined above yields the components of $\boldsymbol{\Delta}$ ($\Delta p_1 \dots \Delta p_K$) that minimize the cost function $S(\boldsymbol{\Delta})$ by way of weighted least squares. Thus, we obtain corrections to each parameter, $p_k \mapsto p_k + \Delta p_k$.

There are, however, important caveats for differential corrections.

1. The DC method is an iterative procedure because every step in the parameter space will result in a new location of that parameter space where the circumstances (i.e., local derivatives) have changed. The nonlinearity of the hyperspace determines the maximum step size we can take for each parameter. The method has no way to control the step size; if all Δp_k are smaller than the typical distance in the parameter space over which the

derivatives do not change substantially, i.e., where the local subspace is largely linear, the method will likely converge. However, if the step size is so large that nonlinearity starts to dominate, the method will likely diverge.
2. Parameter adjustment is unconstrained. This implies that the method can give unphysical results even when convergence is achieved.
3. Missing higher-order derivatives. Equation (8.8) is a *linearized* version of the fully expanded Taylor series. This means that the computation of the cost function does not *sense* their influence, while parameter correlations do. The divergence or, even worse, false convergence may occur when the two conditions (the lack of higher-order differences and parameter correlation) interact (Wilson & Van Hamme 2003).

Pioneering work on introducing DC to the field of eclipsing binary stars was done by Wyse & Kron (1939) and further facilitated by Irwin (1947) with his least-squares lookup tables; first, widespread computer implementation for eclipsing binaries was done by Wilson & Devinney (1971) in their groundbreaking code `WD`.

Two methods for improving convergence stability in `WD` have been proposed: the method of multiple subsets (Wilson & Biermann 1976) and the Levenberg–Marquardt algorithm (Levenberg 1944; Marquardt 1963).

8.2.1 Method of Multiple Subsets

Failing convergence due to nonlinearity of the problem and parameter correlations may be remedied to some extent by breaking the full parameter set into several subsets and thus reducing the complexity of the correlations in a given iteration (Wilson & Biermann 1976). We refer to this as the method of multiple subsets.

Let us assume that the full parameter set consists of six parameters, p_1 through p_6, and that the correlation between them is so strong that it causes bad convergence behavior. The method of multiple subsets breaks this set into two or more subsets, i.e., $A = (p_1, p_2, p_3)$ and $B = (p_4, p_5, p_6)$. Subsets A and B are submitted to the minimizer separately, where each subsequent run is based on the results of the previous run:

$$AB \rightarrow \text{failure}; \quad A \rightarrow B \rightarrow A \rightarrow B \rightarrow \cdots \rightarrow \text{success}, \tag{8.14}$$

with the output from A being input for B and vice versa. This method shows a significant convergence improvement whenever the correlation between parameters is strong. The drawback to the method of multiple subsets is prolongation of the computation time and the underestimation of standard errors of the determined solution. The latter may be corrected by submitting the whole set to the minimizer at the very end of the minimization process, but only for reading out standard errors from the covariance matrix.

8.2.2 Levenberg–Marquardt Algorithm

The Levenberg–Marquardt algorithm (Levenberg 1944; Marquardt 1963) is one of the most widely used and appreciated numerical methods for function minimization.

It has become a standard technique for nonlinear least-squares problems and can be thought of as a combination of the steepest descent method and the Gauss–Newton method (Press et al. 2007). The idea is to replace the equation for steps $\boldsymbol{\Delta}$ (Equation (8.13)) with the damped version:

$$\boldsymbol{\Delta} = (\boldsymbol{A}^T \boldsymbol{W} \boldsymbol{A} + \lambda \ \text{diag}\,[\boldsymbol{A}^T \boldsymbol{W} \boldsymbol{A}])^{-1} \boldsymbol{A}^T \boldsymbol{W} \boldsymbol{\xi}, \tag{8.15}$$

where λ is the (non-negative) damping parameter. The original Levenberg version used a slightly different damping term ($+\lambda \boldsymbol{I}$), but such damping was indiscriminate of the nonlinearity. The Marquardt form (Equation (8.15)) is superior because it "senses" the curvature of the parameter space encoded in the term $\boldsymbol{A}^T \boldsymbol{W} \boldsymbol{A}$, so it can take larger steps along smaller gradients and smaller steps along larger gradients.

To start the minimization, we provide an initial guess for the parameter vector $\boldsymbol{p}$. When the topology of the parameter space is relatively smooth and has an obvious minimum, any "ballpark" guess will work fine; in the case of a more complicated topology with many shallow minima, the algorithm will converge only if the initial guess is already close to the final solution.

The chosen value of λ depends on the convergence behavior. Smaller values of λ reduce the algorithm to the least-squares (Gauss–Newton) method, while larger values of λ turn the algorithm into steepest descent. Once the parameter correction is under a certain threshold or the cost function does not change appreciably, the latest parameter vector $\boldsymbol{p}$ is adopted and iteration is stopped.

The Levenberg–Marquardt algorithm was first applied to eclipsing binaries by Đurašević (1992) and later by Hill & Rucinski (1993). Now all major modeling codes, including WD, support it.

8.3 Powell's Direction Set Method

Often neglected and overlooked, Powell's direction set method[1] (Brent 1973, Chapter 7) is a derivativeless multidimensional method that utilizes a 1D minimization algorithm along the direction in parameter hyperspace. It is quadratically convergent, which makes it one of the fastest methods for solving nonlinear minimization problems. It was first applied to eclipsing binaries by Prša & Zwitter (2007).

The basic idea of Powell's direction set method is to select a starting point $\boldsymbol{P}(p_1, \ldots, p_n)$ in the parameter hyperspace and minimize the cost function along a chosen direction vector $\boldsymbol{n}$ using a 1D minimizer such as bracketing or Brent's parabolic method (Brent 1973, Chapter 5). There is a wealth of direction set method derivatives that differ only on how and at which point of iteration the next direction vector $\boldsymbol{n}$ is chosen (Press et al. 2007). The most appreciated scheme for choosing successive directions is the conjugate gradient method, which we briefly describe here.

When the cost function is minimized along the direction $\boldsymbol{n}$, the gradient in the obtained minimum is necessarily perpendicular to that direction. If it were not, that

[1] Although the method is still referred to as Powell's method, the algorithm used most frequently was developed by Brent (1973), who pointed out and corrected several serious flaws of Powell's original algorithm.

would mean that the projection of the gradient to that direction is non-null, which would in turn mean that the point was not really a minimum. The cost function may be expanded in a Taylor series around that minimum:

$$\begin{aligned} f(\boldsymbol{p}) &= f(\boldsymbol{P}) + \sum_i \frac{\partial f(\boldsymbol{P})}{\partial p_i} p_i + \frac{1}{2} \sum_i \sum_j \frac{\partial^2 f(\boldsymbol{P})}{\partial p_i \partial p_j} p_i p_j + \ldots \\ &= f(\boldsymbol{P}) - \boldsymbol{b} \cdot \boldsymbol{p} + \frac{1}{2} \boldsymbol{p}^T \boldsymbol{H} \boldsymbol{p}, \end{aligned} \tag{8.16}$$

where $\boldsymbol{b}$ is the negative gradient of f at point $\boldsymbol{P}$ and $\boldsymbol{H}$ is the Hessian matrix of second partial derivatives of f at $\boldsymbol{P}$. The gradient is then simply expressed as:

$$\nabla f(\boldsymbol{p}) = \boldsymbol{H}\boldsymbol{p} - \boldsymbol{b}. \tag{8.17}$$

From Equation (8.17) we may immediately deduce the change of the gradient along the chosen direction:

$$\delta(\nabla f(\boldsymbol{p})) = \delta(\boldsymbol{H}\boldsymbol{p} - \boldsymbol{b}) = \boldsymbol{H}(\delta \boldsymbol{p}). \tag{8.18}$$

To adopt the best possible direction from the found minimum $\boldsymbol{P}$, the method must seek an orthogonal (conjugate) direction to the former direction vector, which has to point along the direction of the gradient. If the former direction is denoted with $\boldsymbol{n}$ and the new direction is denoted with $\boldsymbol{m}$, then:

$$\boldsymbol{n} \cdot \delta(\nabla f(\boldsymbol{p})) = \boldsymbol{n}^T \boldsymbol{H} \boldsymbol{m} = 0. \tag{8.19}$$

When Equation (8.19) holds for vectors $\boldsymbol{n}$ and $\boldsymbol{m}$, they are said to be *conjugate*. For as long as the minimization is done *only* in conjugate directions, a single minimization along a given direction is necessary, which implies quadratic convergence. Note that the gradients need not be calculated at any point; we only use the orthogonality implication. This enables Powell's direction set method to preserve a derivativeless nature.

8.4 Nelder & Mead's Downhill Simplex Method

As we have seen, Powell's direction set method (and most other direct multidimensional minimization methods) uses sequential 1D minimizers to achieve convergence. The notable exception is Nelder & Mead's downhill simplex method (Nelder & Mead 1965), which gives remarkable results when the parameter hyperspace is flat and correlations between parameters are strong (Press et al. 2007). The method requires only function evaluations, not derivatives; it is thus not very efficient in terms of the number of evaluations that it requires, but its *generality* pays off for solving difficult nonlinear problems where other methods fail.

The simplex method acts in n-dimensional parameter hyperspace. It constructs n vectors $\boldsymbol{p}_i$ from the vector of initial parameter values $\boldsymbol{x}$ and the vector of step sizes $\boldsymbol{s}$ as follows:

$$\boldsymbol{p}_i = (x_0, x_1, \ldots, x_{i-1}, x_i + s_i, x_{i+1}, \ldots, x_n). \tag{8.20}$$

These vectors form $(n + 1)$ vertices of an n-dimensional simplex. During each iteration, the algorithm attempts to improve parameter vectors $\boldsymbol{p}_i$ by modifying the vertex with the highest function value by simple geometrical transformations (Figure 8.1): reflection, reflection followed by expansion, contraction, and multiple contraction (Galassi et al. 2003). Using these transformations, the simplex moves through parameter space toward the closest minimum, where it contracts itself.

The basic form of the simplex method was first applied to eclipsing binaries by Kallrath & Linnell (1987).

8.5 Adaptive Simulated Annealing

A numerical method for finding a *global* (in contrast to closest) minimum that is gaining on use is derived from classical simulated annealing (Metropolis et al. 1953) and Monte Carlo methods and is called adaptive simulated annealing (Ingber 1989, 1996).

The general idea of all simulated annealing algorithms is to minimize the cost function according to the three-step procedure:

- Given a state $\boldsymbol{p}$ (a set $\boldsymbol{p}$ of parameters p_i), a change is imposed to each parameter according to some prescribed probability distribution function $G(y_i, T_i)$, where $y_i \in [-1, 1]$ is a random variable and T_i is a simulated temperature that determines the amount of the displacement.
- If the imposed change decreases the cost function, the new state will be accepted. If the imposed change increases the cost function, a new state will be rejected *or* accepted, depending on the probability set by the annealing "temperature".
- A simulated temperature parameter is decreased by the amount that is governed by the annealing scheduling function, and the state is submitted to the next iteration.

Such a logic enables the method to escape from local minima, as even if the solution lies within a local minimum, there is still a chance that the algorithm will displace it.

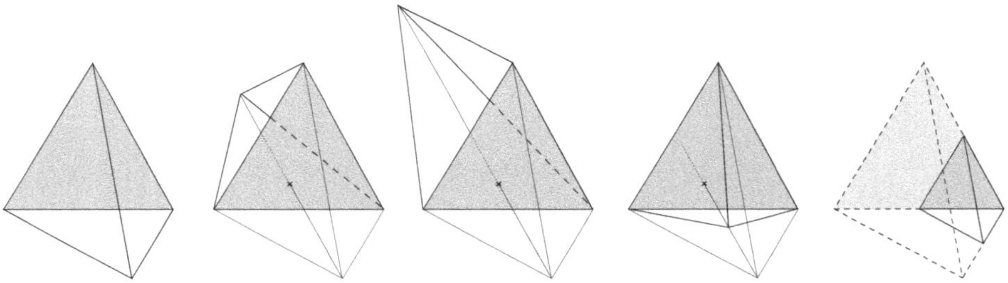

Figure 8.1. Four possible transformations of the 3D simplex: reflection, reflection followed by expansion, contraction, and multiple contraction. The vertex being transformed is the one with the highest functional value. Using these transformations, the simplex moves through the parameter hyperspace toward the closest minimum, where it contracts. At the end of each iteration the vertex with the lowest functional value is returned.

Adaptive simulated annealing differs only a little from this general scheme, by providing hyperspace-dependent probability distribution functions and annealing scheduling. This adapts the algorithm to the problem at hand automatically.

A serious downside to any simulated annealing or similar algorithm is the computational time cost. In order to preserve statistical properties that guarantee the method to find the global minimum and to remove the ambiguity in the locality of other minima, temperature annealing has to be done slowly enough.

Simulated annealing has been applied to eclipsing binaries by Nightfall, a freely available code for modeling eclipsing binary stars, available from http://www.hs.uni-hamburg.de/DE/Ins/Per/Wichmann/Nightfall.html.

8.6 Formal Errors of the Fit

Irrespective of the minimizing method used, the χ^2 cost function features a ready way to estimate errors that correspond to the determined parameters and to examine the shape of the hyperspace around the minimum. To see how this is done, consider again Equation (8.10) expressed in terms of parameters rather than parameter corrections:

$$\chi^2(\boldsymbol{p}) = (\boldsymbol{F}^{\text{obs}} - \boldsymbol{A}\boldsymbol{p})^T \boldsymbol{W}(\boldsymbol{F}^{\text{obs}} - \boldsymbol{A}\boldsymbol{p}), \tag{8.21}$$

where $\boldsymbol{p}$ is the vector of all parameters of the model and $[A]_{jk} = \partial F_j^{\text{syn}}/\partial p_k$. The χ^2 function is taken to be smooth, with finite second derivatives across the definition range. As before, we obtain an expression for $\boldsymbol{p}$ by requiring that $\partial\chi^2(\boldsymbol{p})/\partial\boldsymbol{p} \equiv \nabla_p\chi^2(\boldsymbol{p}) = 0$:

$$\boldsymbol{A}^T\boldsymbol{W}(\boldsymbol{F}^{\text{obs}} - \boldsymbol{A}\boldsymbol{p}) = 0 \quad \Rightarrow \quad \boldsymbol{p} = (\boldsymbol{A}^T\boldsymbol{W}\boldsymbol{A})^{-1}\boldsymbol{A}^T\boldsymbol{W}\boldsymbol{F}^{\text{obs}}, \tag{8.22}$$

where $(\boldsymbol{A}^{\text{T}}\boldsymbol{W}\boldsymbol{A})^{-1} \equiv \boldsymbol{V}$ is the covariance matrix (Avery 1991):

$$[V]_{ij} = \langle(p_i - \bar{p}_i)(p_j - \bar{p}_j)\rangle = \langle\delta p_i \delta p_j\rangle. \tag{8.23}$$

Angle brackets $\langle\rangle$ refer to a weighted average taken over all possible values of the enclosed expression. From Equation (8.23) it follows that the diagonal elements of the covariance matrix are standard deviations σ_{p_i}; the off-diagonal elements correspond to the correlation coefficients:

$$[V]_{ij} = [\rho]_{ij}\sigma_{p_i}\sigma_{p_j}, \tag{8.24}$$

where $\boldsymbol{\rho}$ is the correlation matrix and ρ_{ij} are the correlation coefficients.

To describe the shape of the hyperspace in the vicinity of the minimum, Equation (8.21) is expanded in Taylor series up to the quadratic term:

$$\begin{aligned} \chi^2(\boldsymbol{p}) &= \chi^2(\boldsymbol{p}_0) + \sum_k \frac{\partial\chi^2}{\partial p_k}(p_k - p_{k,0}) \\ &+ \frac{1}{2}\sum_{i,j}\frac{\partial^2\chi^2}{\partial p_k \partial p_l}(p_k - p_{k,0})(p_l - p_{l,0}) \\ &\equiv \chi^2(\boldsymbol{p}_0) - \boldsymbol{b}\cdot\boldsymbol{p} + \boldsymbol{p}^T\boldsymbol{H}\boldsymbol{p}, \end{aligned} \tag{8.25}$$

where $\boldsymbol{b} = -\nabla\chi^2(\boldsymbol{p})$ and $\boldsymbol{H}$ is the Hessian matrix of second derivatives. The gradient of χ^2 may be written out explicitly:

$$\frac{\partial\chi^2}{\partial p_k} = -2\sum_i \frac{[F_i^{\text{obs}} - F_i^{\text{syn}}(\boldsymbol{p})]}{\sigma_i^2}\frac{\partial F_i^{\text{syn}}(\boldsymbol{p})}{\partial p_k}, \tag{8.26}$$

where index k goes over all parameters of the model. Evaluating the second derivative yields:

$$\frac{\partial^2\chi^2}{\partial p_k \partial p_l} = 2\sum_i \frac{1}{\sigma_i^2}\left[\frac{\partial F_i^{\text{syn}}(\boldsymbol{p})}{\partial p_k}\frac{\partial F_i^{\text{syn}}(\boldsymbol{p})}{\partial p_l} - (F_i^{\text{obs}} - F_i^{\text{syn}}(\boldsymbol{p}))\frac{\partial^2 F_i^{\text{syn}}(\boldsymbol{p})}{\partial p_l \partial p_k}\right]. \tag{8.27}$$

The following test can be performed for a nondegenerate critical point $\boldsymbol{P}$, in which the gradient of χ^2 is zero: if the Hessian is positive definite at $\boldsymbol{P}$, then χ^2 attains a local minimum at $\boldsymbol{P}$. If the Hessian is negative definite at $\boldsymbol{P}$, then χ^2 attains a local maximum at $\boldsymbol{P}$. If the Hessian has both positive and negative eigenvalues, then $\boldsymbol{P}$ is a saddle point for χ^2.

Unfortunately, this test cannot qualify the minima across the entire hyperspace—it is confined to the given local minimum. Furthermore, the test is often rendered inconclusive because of the noise and parameter correlations; Table 8.1 lists coefficients of the Hessian matrix for a synthetically created light curve with Poissonian scatter of 0.02 mag, evaluated at the global minimum. As we have stressed in the opening of this chapter, we need diverse data sets, as parameter sensitivity varies with the types of data. Inspecting the coefficients of the Hessian matrix highlights this point: parameters q and a make the matrix singular because both parameters do not affect the shape of a light curve. Had we also used an RV curve, the Hessian would have been different. The Hessian test can thus help the analysis of the shape of the local minimum, but it cannot aid in handling the degeneracies in the hyperspace.

Table 8.1. Example of the Hessian Matrix

	i	Ω_1	Ω_2	T_{eff1}	T_{eff2}	q	a
i	0.321155	−0.003557	−0.003371	0.003958	0.021218	0.000576	0.000002
Ω_1	−0.003557	0.029753	−0.002312	1.712314	−1.842537	−0.000985	−0.000009
Ω_2	−0.003371	−0.002312	0.028974	−1.637029	1.524242	−0.003917	−0.000009
T_{eff1}	0.003958	1.712314	−1.637029	0.001368	−0.001186	0.000000	0.000000
T_{eff2}	0.021218	−1.842537	1.524242	−0.001186	0.000950	−0.000000	−0.000000
q	0.000576	−0.000985	−0.003917	0.000000	−0.000000	0.015099	0.000042
a	0.000002	−0.000009	−0.000009	0.000000	−0.000000	0.000042	−0.000000

Notes. The model consists of a single light curve that is computed synthetically with Poissonian scatter 0.02 mag. The Hessian is evaluated at the global minimum. The last two parameters, mass ratio q and semimajor axis a, make the Hessian singular; this is expected, because the shape of the light curve is insensitive to the value of q or a. To identify such parameters, a singular value decomposition method can be employed (Galassi et al. 2003).

8.7 Handling Degeneracies Efficiently

We have already stressed how important it is to discern between local and global minima and to guide the minimizer toward the deepest minimum. The degeneracy that is inherent to the problem due to nonlinearity and parameter correlations needs to be analyzed *individually for every modeled data set* if error estimates and solution reliability are to be assessed rigorously.

One of the common misconceptions about degeneracy is that it originates in data signal. Researchers sometimes agonize over deciding among adjacent minima in the neighborhood of the optimal solution in an effort to determine the "best-fit" solution. This is not only futile but also leads to a (frequently severe) underestimation of formal errors of the fit. Discerning local minima would suggest that the model itself is able to discern between *physically indiscernible* minima: the shape and layout of the parameter hyperspace within minima regions are *not* determined by the data signal, but by the data noise. If a single observed data point changed, or if the error estimates were slightly different, the microtopology of the hyperspace would also change. Yet the distinction between these solutions is physically irrelevant—whenever a solution changes by an amount less than the level of accuracy of the data, individual minima within are *all* equivalent. The co-dimension is defined as the number of mutually independent directions in which the gradient of the cost function is 0. This may be easily visualized in the 3D case: a minimum may be either a point, as the lowest part of the sinclination, a line, as a V-shaped canyon, or a region, as a flat bottom of the valley. The co-dimension of the minimum is always larger than 1 and can span up to the dimension of the hyperspace itself: the minima are bound regions rather than points, within which all solutions are equivalent. This constitutes a physical degeneracy of the model that can be addressed only by acquiring more data.

Another common misconception is to try to extract physical parameters of the model from the data that are not sensitive to that parameter. A good example is trying to obtain the semimajor axis a from photometric light curves; another such example is trying to discern between two symmetric values of inclination with respect to 90°, i.e., 80° and 100°. Since such cases do not contain any physical information (other than the fact that there *is* no physical content in the given observed data set), they do not constitute a physical degeneracy.

Apart from those two cases, any other effect that causes the value of the cost function to be (roughly) the same for different sets of parameters constitutes a physical degeneracy that we must learn to recognize and handle properly. An example of a true (and often present) degeneracy is between the inclination i and surface potentials Ω_1 and Ω_2, depicted in Figure 8.2: smaller values of inclination may be compensated by increasing the radius of the star, which corresponds to lower values of Ω_1. This degeneracy is difficult to handle without additional data sources or external conditional constraints (i.e., the ones imposed by a human, not by the model).

To be able to tackle such problems, heuristic approaches have been developed and incorporated in eclipsing binary models: heuristic scanning and parameter kicking (Prša & Zwitter 2005).

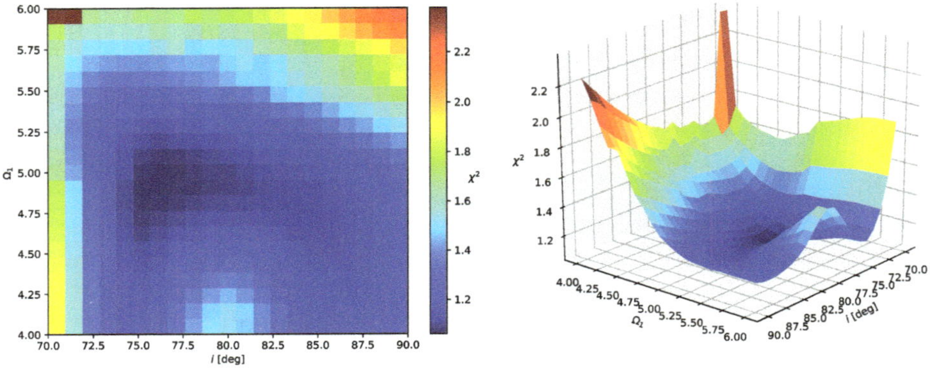

Figure 8.2. Degeneracy between the inclination i and the surface potential of the primary star Ω_1. Left: χ^2 topology of the i–Ω_1 cross section. Right: 3D plot of the i–Ω_1 cross section. Color-coding corresponds to the value of the cost function. The plots were computed for a simulated binary with two light curves with Poissonian scatter 0.02 mag. The degeneracy may be easily understood: lower inclinations may be compensated by enlarging the radius of the primary star.

8.7.1 Heuristic Scanning

All minimization algorithms, even the simplex method with its property of guaranteed convergence for as long as the minimum is enclosed in the initial simplex, can be stuck in a local minimum, particularly since parameter hyperspace in the vicinity of the global minimum is typically very flat, with lots of local minima. In addition, the global minimum may be shadowed by data noise and degeneracy.

Heuristic scanning is a brute-force method run on top of any minimization algorithm we have discussed so far. It selects a set of starting points in the parameter hyperspace and starts the minimization from each such point. The points may be gridded, stochastically dispersed, distributed according to some probability distribution function, etc. The algorithm then sorts all solutions by the cost function and weighs the obtained parameter values accordingly: heuristic runs with the smallest values of the cost function correspond to the deepest minima and should thus have a higher weight—they are the most likely candidates for the global minimum.

The weighted values of the adjusted parameters are then put into histograms, from which the mean and standard deviation of parameter values can be calculated. These estimates are thus based on the ensemble statistic, as they do not depend on the formal errors of the numerical method. Figure 8.3 shows an example of such histograms for the effective temperature ratio $\tau = T_2/T_1$, adopted from Prša & Zwitter (2005). The heuristic scanning results for this particular example are largely insensitive to the observed data uncertainties: for three significantly different cases (labeled *best*, *medium*, and *worst* in Figure 8.3), the outcome of the histogram fit is approximately the same. Histograms for other parameters have somewhat larger standard deviations because of the degeneracies. Note that reliable statistics implies many starting points for heuristic scanning, which in turn implies a significant impact on computation time: each additional scan linearly contributes to the time cost. This method is a variant of importance sampling (Serfozo 2009).

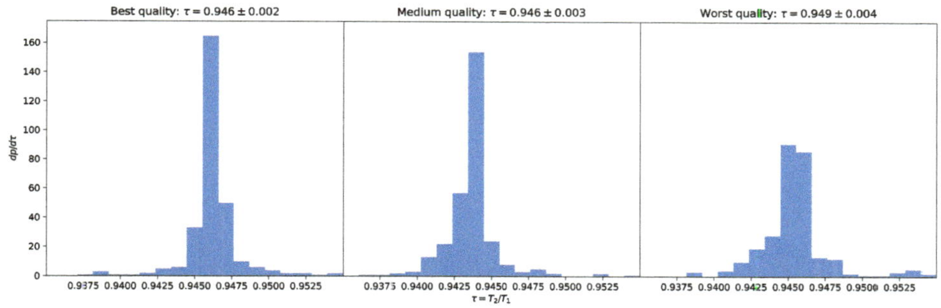

Figure 8.3. Posterior temperature ratio distributions obtained by heuristic scanning. The plots show $\tau = T_2/T_1$ histograms for the same data set and three levels of noise: $\sigma_{\rm LC} = 0.005$, $\sigma_{\rm RV} = 5$ kms^{-1} (left), $\sigma_{\rm LC} = 0.015$, $\sigma_{\rm RV} = 15$ kms^{-1} (middle), and $\sigma_{\rm LC} = 0.025$, $\sigma_{\rm RV} = 25$ kms^{-1} (right). Heuristic scanning is largely insensitive to data precision as long as there are sufficient data points to determine both eclipse depths. The obtained values of temperature ratios are statistical and may be compared to the true value of $\tau = 0.9452$ that was used to create the data set. Adopted from Prša & Zwitter (2005). © 2005. The American Astronomical Society.

Because of the data noise and degeneracy, the global minimum is essentially never a single point (with its corresponding uncertainty); it is actually a *region* (with its corresponding uncertainty) in the parameter hyperspace. Such a region encompasses many adjacent minima, the depths of which are physically indistinguishable—a single observed data point with its standard deviation may change the topology within that region. To identify these regions, we compute *convergence tracers*—2D cross sections of the parameter hyperspace, tracing parameter values from each starting point, iteration after iteration, all the way to the converged solution. Attractors—regions that attract most convergence traces—within these cross sections reveal parameter correlations and degeneracy. Inspecting such convergence tracers provides additional insight into the quality and integrity of the solution. Local minima are those that lie outside of the deepest attractor(s); those are the ones that need to be identified and avoided. Figure 8.5(a) shows the i–Ω_1 convergence tracer computed for the same simulated data as in Figure 8.2. The correlation between i and Ω_1 is evidently very flat at inclinations close to 90°, which may be easily understood: the model is able to compensate for smaller inclinations by enlarging the radius of the star and vice versa. Therefore, we should not trust light-curve analysis to disentangle these parameters by itself—additional constraints are needed. This issue will be further discussed in the next section.

8.7.2 Parameter Kicking

Another possible approach to detect and escape from local minima is to use a stochastic method such as simulated annealing. However, such methods are notoriously slow. Thus, instead of a full-featured simulated annealing scan, a simple new procedure has been developed that achieves the same effect as stochastic methods, but in a significantly shorter time. The idea is as follows: whenever a minimum is reached within a given fractional accuracy, the algorithm runs a globality assessment on that minimum.

Parameter kicking is a way of knocking the obtained parameter set out of the minimum: using a Gaussian probability distribution function, the method randomly picks an offset for each parameter. The strength of the kick is determined by the Gaussian dispersion σ_{kick}, which depends on the minimum globality assessment parameter χ^2_ν, Equation (8.5). If χ^2_ν is high, then the kick should be strong, but if it is low, i.e., around $\chi^2_\nu \sim 1$, then only subtle perturbations should be allowed. Tests have shown that a simple expression such as:

$$\sigma^2_{\text{kick}} = \frac{0.5\chi^2_\nu}{100} \tag{8.28}$$

works very efficiently in the case of partial eclipses (Prša & Zwitter 2005). This causes σ^2_{kick} to assume a value of 0.5 for 10σ offsets and 0.005 for 1σ offsets, being linear in between. Note that this σ_{kick} is *relative*, i.e., given by:

$$\sigma^{\text{abs}}_{\text{kick}} = x\,\sigma^{\text{rel}}_{\text{kick}}, \tag{8.29}$$

where x is the value of the given parameter. When convergence within the given fractional accuracy is reached, parameters are kicked with respect to the depth of the minimum and minimization is restarted from displaced points. The influence of

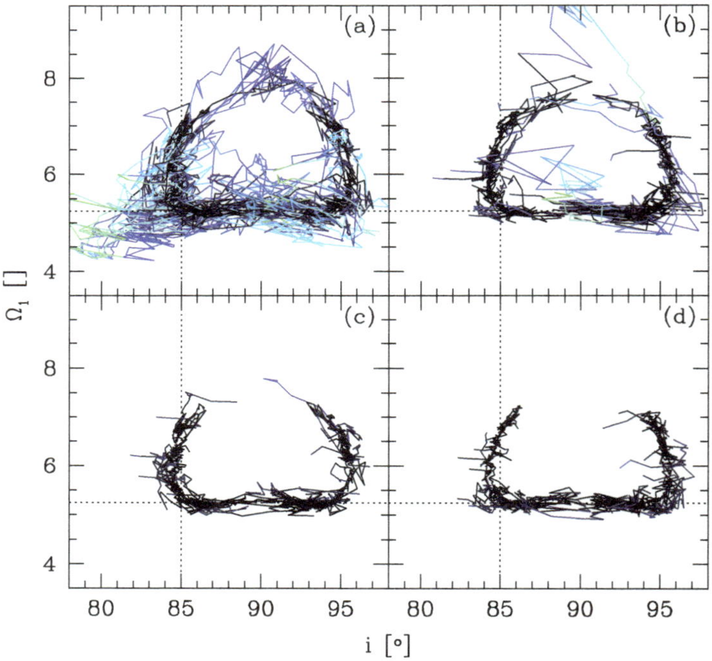

Figure 8.4. Convergence tracer for the i–Ω_1 cross section. Colors denote the value of the cost function: the darker the trace, the lower the value of χ^2. This particular case is a notorious example of a very difficult to handle correlation between the inclination and effective potentials (proxies for the radii) of both components (only Ω_1 correlation is depicted for brevity). Individual plots denoted with letters (a) through (d) show the result of simplex heuristic scanning from zero to three consecutive parameter kicks. Cross-hairs mark the position of the true minimum. Attractors are symmetric to $i = 90°$ but still very flat at the $i \sim 85°$–$90°$ interval, which means that the obtained simplex solution should not be blindly trusted; rather, additional constraining might be needed.

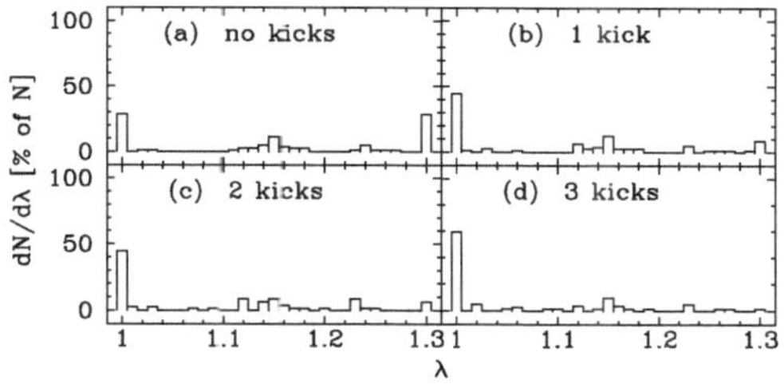

Figure 8.5. χ^2_ν-histogram for the initial heuristic scan and three consecutive parameter kicks. The success of parameter kicking is evident, as after only three consecutive kicks the percentage of scans that converge within 1% of the optimal value of χ^2_ν is *doubled* from ~30% to ~60% (Prša & Zwitter 2005). Parameter kicking proves to be even more efficient in the case of exclusively photometric observations, with improvement from ~15% to ~75% in convergence after three consecutive kicks (Prša & Zwitter 2005). © 2005. The American Astronomical Society.

consecutive parameter kicking with the simplex method is depicted in Figure 8.4; out of all heuristic traces only ~30% initially converge to within 1% of the optimal value of χ^2_ν, whereas this percentage steadily grows to ~60% after three kicks. Note that the actual value of χ^2_ν is not important; having under- or overestimated the σ_k values, the value of χ^2_ν will be less or more than 1. It is the relative offset from the minimal value of χ^2_ν that is important. Figures 8.5(b)–(d) show a significant improvement to the solution introduced by these consecutive kicks. Parameter kicking is able to quickly escape from local minima and thus rapidly increase convergence efficiency of the simplex method. The downside of parameter kicking is the time cost: each additional kick linearly adds to the overall execution time.

8.8 Markov Chain Monte Carlo Samplers

Arguably the best heuristic approach to fitting the data is by using the Markov chain Monte Carlo (MCMC) samplers, but you better have a high-performance computing cluster at your disposal. They are used to sample probability distribution functions (pdf's) of model parameters. If $\mathsf{p}(\boldsymbol{p})$ is the prior pdf for parameters $\boldsymbol{p}$, the posterior pdf given observations D is given by the Bayes rule:

$$\mathsf{p}(\boldsymbol{p}|D) = \frac{\mathsf{p}(D|\boldsymbol{p})\mathsf{p}(\boldsymbol{p})}{\mathsf{p}(D)}, \tag{8.30}$$

where the constant $\mathsf{p}(D)$ is called *evidence* and $\mathsf{p}(D|\boldsymbol{p})$ is the likelihood function. Evidence is particularly difficult to compute, as it encompasses all the complexity of the priors and of the likelihood function, but as it serves only as a scaling constant, we do not have to compute it to get the posterior pdf. Moreover, we have seen already that heuristic scanning was insensitive to the normalization of the cost function, and so is MCMC; thus, computing evidence is not necessary to infer parameter posteriors.

As a heuristic method, MCMC relies only on function evaluations. Thus, in terms of complexity, it is no different from heuristic scanning, but it is much more powerful insofar that convergence is driven rather than derived from random sampling of the parameter space.

It is important to clarify the probabilistic description of parameter values inherent to inference. There are many great books and monographs written on the topic (e.g., MacKay 2002; Rasmussen & Williams 2005; Hogg et al. 2010; Hogg 2012; Hogg & Foreman-Mackey 2017) that cover the ins and outs of probabilistic inference in due detail. Here we only scratch the surface in order to justify the approach and elucidate the results of MCMC-powered sampling. Given the prior pdf $\mathsf{p}(\boldsymbol{p})$, we can compute expectation values $E(f(\boldsymbol{p}))$ for any function $f(\boldsymbol{p})$ as:

$$E(f(\boldsymbol{p})) = \int f(\boldsymbol{p})\mathsf{p}(\boldsymbol{p})d\boldsymbol{p}, \tag{8.31}$$

where we integrate over the entire definition range of $\boldsymbol{p}$. These expectation values are simple means of $f(\boldsymbol{p})$ given the prior pdf $\mathsf{p}(\boldsymbol{p})$. The point of MCMC sampling is to approximate integral (8.31) with the sum:

$$E(f(\boldsymbol{p})) = \frac{1}{N}\sum_{i=1}^{N} f(\boldsymbol{p}_i). \tag{8.32}$$

The sampler requires two inputs: the probability function $f(\boldsymbol{p})$, which is a product of the likelihood function $\mathsf{p}(D|\boldsymbol{p})$ and the prior pdf $\mathsf{p}(\boldsymbol{p})$, and a set of priors $\mathsf{p}(\boldsymbol{p})$, i.e., any prior knowledge we have on the values of the parameters. A good sampler needs to be able to traverse the parameter space in a *fair* way, which implies being able to draw a new state $\boldsymbol{p}_{i+1}$ from the previous state $\boldsymbol{p}_i$. In its simplest form, the sampler does the following:

- Draw a new state $\boldsymbol{p}_{i+1}$. The drawing is done from the proposal pdf, $\pi(\boldsymbol{p}_{i+1}|\boldsymbol{p}_i)$. The proposal pdf can be in principle any generative function, but the typical choice is a multivariate Gaussian centered on $\boldsymbol{p}_i$. The chosen proposal pdf needs to obey the symmetry condition, $\pi(\boldsymbol{p}_{i+1}|\boldsymbol{p}_i) = \pi(\boldsymbol{p}_i|\boldsymbol{p}_{i+1})$.
- Draw r from the uniform distribution $\mathcal{U}(0, 1)$. This auxiliary parameter will drive convergence toward a stationary solution.
- Adopt or reject the new state $\boldsymbol{p}_{i+1}$. If $f(\boldsymbol{p}_{i+1})/f(\boldsymbol{p}_i) > r$, adopt $\boldsymbol{p}_{i+1}$; otherwise, set $\boldsymbol{p}_{i+1} = \boldsymbol{p}_i$. If the dynamical range of the probability function is large (which is almost always the case), it is better to do this step in log space: if $\log f(\boldsymbol{p}_{i+1}) - \log f(\boldsymbol{p}_i) > \log r$, adopt $\boldsymbol{p}_{i+1}$; otherwise, reject it.

This sampler is called the Metropolis MCMC, as its adoption and rejection of proposed states $\boldsymbol{p}_{i+1}$ are motivated by simulated annealing (Metropolis et al. 1953). The chain $\boldsymbol{p}_i$, $i = 1 \dots N$ represents a biased random walk through parameter space in the sense that the number of iterations at a certain state $\boldsymbol{p}_i$ is proportional to the probability function $f(\boldsymbol{p}_i)$ (Hogg & Foreman-Mackey 2017).

Except for the simplest of problems, it is essentially impossible to know when to stop iterating; while the condition can be described simply as "when the posterior

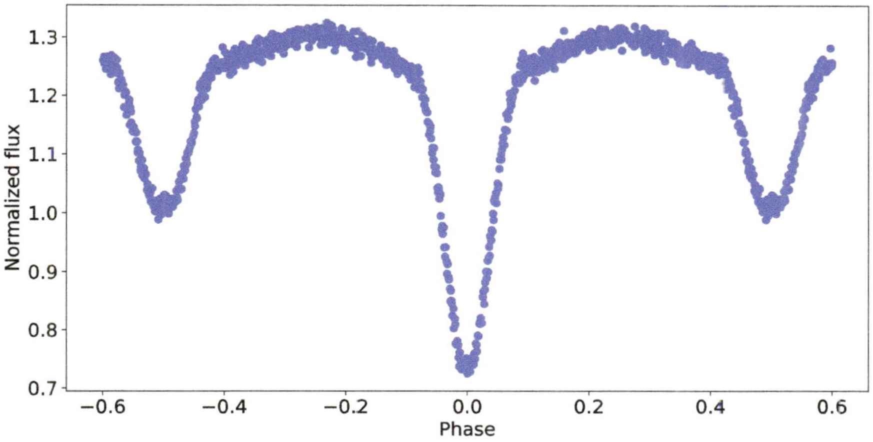

Figure 8.6. Synthetic light curve of a close eclipsing binary used in the test of MCMC sampling. The light curve corresponds to a canonical binary star observed in the Johnson V passband; the light curve was generated using the model, and the only modification made was the addition of purely white noise. As such, the light curve is woefully idealized compared to what we would encounter in real life, but it serves the purpose of stressing the benefits of probabilistic inference.

pdf has been sampled adequately," in reality this is impossible to gauge. Instead, we will be satisfied when (large) subsets of the chains exhibit the same (within reasonable errors) posterior pdf, implying that each chain traversed the most probable part of the parameter space several (many) times. A particularly useful diagnostic of convergence, proposed by Gelman & Rubin (1992), compares the variance in parameters within a chain to the variance across many heuristic chains (i.e., chains started from different parts of the parameter space; see Section 8.7.1).

Consider the synthetic light curve depicted in Figure 8.6. This is a severely idealized light curve: the generative function is the model itself, the noise is purely white, the orbit is circular, and there are no added complications, such as spots, misalignment, or asynchronous rotation. To demonstrate the overall logic and the process of probabilistic inference to fit the parameters of this light curve, such an idealized light curve serves the purpose perfectly.

We will naively attempt to determine posterior pdf's for the following parameters: semimajor axis a, mass ratio q, center-of-mass velocity v_γ, primary and secondary star surface potentials Ω_1 and Ω_2, inclination i, and secondary star effective temperature T_{eff2}. We say "naively" because, given that the only data set we use here is a single photometric light curve, we do not really expect to be able to fit a and v_γ at all, and q only marginally, from the signature of ellipsoidal deformation. Table 8.2 provides the true values of parameters along with the "top-hat" prior pdf's (1 on the given interval, 0 otherwise).

The iterative Markov chain sampling is done with 128 "walkers." Each walker starts at a unique parameter state $\boldsymbol{p}_k$, $k = 1 \ldots 128$ that is initialized by drawing from the prior pdf. For simplicity, the priors are simple bound intervals, sometimes called uninformative priors. The algorithm then forms Markov chains for each walker,

Table 8.2. Principal Parameters of the Synthetic Binary That Produces the Light Curve Depicted in Figure 8.6

Parameter	Unit	System		Prior pdf
		Primary	Secondary	
a	(RSun)	3.8		[3.0, 5.0]
q		0.8		[0.5, 1.0]
v_γ	(km s^{-1})	−15		[−30, 10]
i	(deg)	85		[80, 90]
e		0		Not sampled
$T_{\rm eff}$	(K)	6500	5500	[4500, 7000]
Ω		4.0	5.0	[2, 10]
A (albedo)		0.6	0.6	Not sampled
β		0.32	0.32	Not sampled
L_V		13.5	11.0	Minimized*
LD model		log	log	
$x_{\rm LD}$		0.73	0.77	Not sampled
$y_{\rm LD}$		0.19	0.15	Not sampled

*Values were computed using least squares.

based on the logic outlined above. As the walkers crawl the parameter space, they traverse sets of parameters and eventually converge to a stationary state. The top panel of Figure 8.7 showcases the convergence where each walker is depicted from its initial state (off-scale because of the large negative value) to its final state after 10,000 iterations. We can see that there are two attractors (and one lost walker that will have converged to the top plateau). The top attractor corresponds to a higher likelihood and, indeed, reflects a global minimum. The bottom attractor corresponds to a lower likelihood and reflects a local minimum.

The bottom panel of Figure 8.7 is rich in content, which we describe in detail now. The depicted samples correspond to the top plateau, with likelihoods between −500 and −450 and a "burn-in" time (the number of skipped steps from the start of the chain) of 200 iterations. The diagonal panels show posterior pdf's of all sampled parameters in blue and fitting Gaussians to posterior pdf's in black. The mean and standard deviation of the Gaussian fit are given on top of each column. The posteriors for the semimajor axis and center-of-mass velocity are flat, which implies that the data are insensitive to these two parameters, as we initially expected. The panels below the diagonal depict two-parameter cross sections. When parameters are not correlated, we expect concentric circles like the ones that correspond to $T_{\rm eff2}$–$\Omega_{1,2}$ or $\mathrm{T}_{\rm eff2}$–i cross sections. When parameters *are* correlated, we expect ellipses, as is the case for the q–$\Omega_{1,2}$ cross sections. Uninformative posteriors fill the entire allowed definition range and result in box-like corellations, such as the a–v_γ cross section. The panels above the diagonal convey the same information in terms of correlation coefficients. Small values indicate no correlation, values close to 1 correspond to strong correlation, and values close to −1 correspond to strong anticorrelation. The cross-hairs depict 1 σ (red) and 2 σ (green) intervals.

It is instructive to compare the true values from Table 8.2 with the posterior pdf means. Given the trivial simplicity of this case, it is not surprising that most parameters fit very well. After all, we used the same model to generate the data and to fit the data, and we only added white noise, both of which are a luxury we never have in real life. All values that are constrained by the light curve are within 1 σ of the true value, but there are several important points to note here:

1. We only sample across a 7D parameter subspace, keeping the remaining parameters at their true values. This further trivializes this exercise and makes the posteriors look exceptionally robust. When dealing with real data, this is not often the case.
2. We mentioned earlier that the mass ratio is only marginally constrained through ellipsoidal variations. This seems to disagree with the information we glean from the posterior pdf for q. But a closer look at the q–$\Omega_{1,2}$ correlations (0.98 and 0.99) immediately reveals that q will be determined correctly only if the surface potentials are determined correctly. The light curve in Figure 8.6 corresponds to a total eclipser, which aids in breaking the degeneracy of the known *sum* of the radii but an unknown *ratio* of the radii (see the $\Omega_1 - \Omega_2$ cross section); our ability to get a reliable posterior pdf for q is thus a result of (a) an idealized case, (b) ellipsoidal variations, and (c) total eclipses.
3. Posterior pdf's critically depend on the assigned uncertainty to data points. If the uncertainties were underestimated, the posteriors would spread into multimodal pdf's. If the uncertainties were overestimated, the width of the posteriors would increase. Neither correctly captures the actual posterior pdf, so great care should be taken to assign realistic uncertainties to the data.
4. Lastly, while we did not evaluate convergence here explicitly, for the reasons mentioned above we *know* that the top plateau converged to the global minimum. In reality it would be better to stop the iteration once the stationary states have been reached and redraw the parameter states for the walkers "stuck" on the bottom plateau from the multivariate Gaussian distribution that is derived from the walkers on the top plateau—or, alternatively, spend eternity in waiting for those walkers to come to their senses and jump over to the highest likelihood state.

One final piece of advice when visualizing the results of the fit is to compute the *average* light curve and standard deviation across all walkers on the top plateau in their last iteration. While this average light curve will not correspond to any particular set of parameters, plotting this average light curve and shading the segment between $\bar{f} - \sigma_{\bar{f}}$ and $\bar{f} + \sigma_{\bar{f}}$ will provide a visual guide to the quality (and time/phase-resolved uncertainty) of the fit.

With this we conclude our discussion on solving the inverse problem, and we urge the reader to commit to learning and using heuristic methods at least in conjunction with, if not in the place of, the traditional deterministic methods.

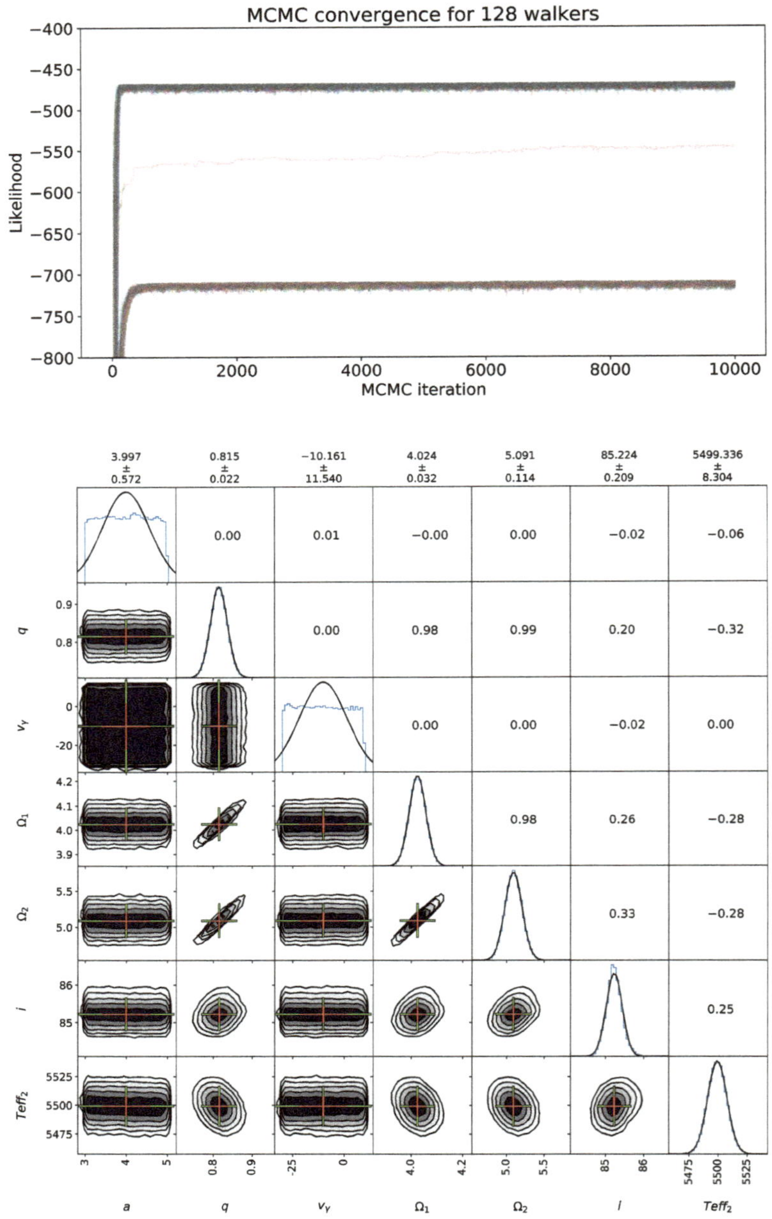

Figure 8.7. Results of the MCMC sampling of the parameter space. Top: dependence of the probability function on iteration for 128 walkers. All walkers (except the one straggler) converge to either a global minimum (top plateau) or a local minimum (bottom plateau). Bottom: triangle plot of the results. The diagonal panels depict posterior pdf's of sampled parameters; the panels below the diagonal visualize parameter cross sections; the panels above the diagonal give the correlation coefficient that corresponds to the parameter cross sections. Each posterior pdf is fitted by a Gaussian. See text for the interpretation of the plot.

8.9 Parameter Constraints as Degeneracy Breakers

In the cases where spectral energy distribution observations are not available, or where they are used only to extract RVs, the degeneracy among parameters may still be so strong that neither heuristics nor probabilistic fitting can break it. In such cases we cannot obtain a satisfactory solution without further constraining the modeled binary.

We discussed in Chapter 3 how the value of the surface potential Ω affects the morphology of the binary. By adopting a semi-detached or contact morphology, we impose a set of physical constraints; for example, both components of a contact system have equal potentials, albedos, gravity-darkening coefficients, etc. If a morphological constraint is not selected appropriately, the model may converge to a physically implausible solution.

We can sometimes make further assumptions. In the case of degeneracy, a solution based on an assumption may be better than having no solution at all. Such an assumption might be the age of the coeval components. Assuming a particular type of the evolutionary track, the luminosities from stellar evolution models may then be obtained (Pols et al. 1995). Another such assumption could be the distance to the binary, i.e., from astrometry (Astraatmadja & Bailer-Jones 2016). Yet another assumption may be that either or both components are main-sequence stars. As a significant percentage of all stars are on the main sequence, there is a fair chance that our assumption might be correct.

Applying a main-sequence constraint to the component(s) of the modeled binary means imposing *M–L–T–R* relations for main-sequence stars (for such relations specific to eclipsing binaries, see, e.g., Torres et al. 2010). In consequence, given a single parameter (e.g., component's effective temperature), all other parameters (its mass, luminosity, and radius) are calculated. This in turn implies that, in the case of circular and nearly circular orbits, the effective potential of the constrained component is fully determined. The main-sequence constraint may be used for testing whether either or both stars may plausibly be main-sequence stars: depending on the behavior of the χ^2 cost function, such a hypothesis may be accepted or rejected.

Additional constraints are not as straightforward as was the case with morphology constraints. For example, by implying the main-sequence constraint, we break the degeneracy by selecting the solution that corresponds to that constraint. It is very important to emphasize that using constraints inappropriately may lead to creating and propagating a circular argument: eclipsing binaries provide absolute parameters for stars, which can then be used to establish various calibrations. The main-sequence constraint uses those calibrations to constrain derived parameters. Should we be tempted to report the values we derived under the main-sequence constraint, we would propagate a circular argument.

Recall from Figure 8.5 that the solution from photometric and RV observations of our test binary indeed suffered from the degeneracy between the inclination and the potentials. If we constrain both components with the main-sequence constraint, potentials Ω_1 and Ω_2 can be calculated, and their values are thus precisely known.

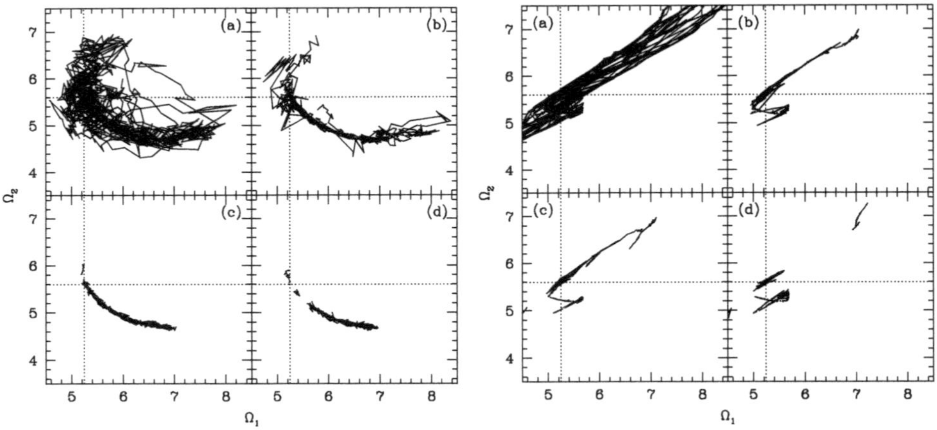

Figure 8.8. Convergence tracers for the Ω_1–Ω_2 cross section without (left) and with (right) the main-sequence constraint imposed on the model. Similar to Figure 8.5, panels (a) through (d) denote successive numbers of kicks (zero to three), and cross-hairs mark the correct solution. Comparing these results immediately shows that the intersection of the two attractors yields the correct solution. Note that there are two intersections because of the model symmetry to the labeling of the two components (primary and secondary roles of both stars may be interchanged).

Any leftover variation in their values is a consequence of the variation in any of the main-sequence parameters (M, L, T, or R) that accommodate different orbital inclinations during the fit. Figure 8.8 shows convergence tracers for a similar heuristic scan to that depicted in Figure 8.5, but this time for the $\Omega_1 - \Omega_2$ cross section without (left) and with (right) the main-sequence constraint imposed on the model. As the main-sequence constraint is very stringent, there is no practical need for heuristic scanning or parameter kicking (both Ω_i values are calculated for the given inclination, and convergence is thus assured from practically *any* point in the hyperspace); the right panel of Figure 8.8 depicts both the heuristic scanning and parameter kicking for comparison purposes. It is evident that both solutions intersect, yielding the correct solution. This is, of course, expected, as our test binary was in fact composed of two main-sequence components.

We would hope that total eclipses reduce the degeneracy, but this also does not necessarily happen. If stars have comparable sizes, the duration of eclipse totality is very short and limb and gravity darkening may obscure its flatness. As geometrical parameters are better constrained in the case of total eclipses (the corresponding hyperspace cross sections feature very narrow valleys), parameter kicking may work to our disadvantage, knocking the solution out of the valley by only a small parameter displacement.

References

Astraatmadja, T. L., & Bailer-Jones, C. A. L. 2016, ApJ, 833, 119

Avery, P. 1991, Applied Fitting Theory I: General Least Squares Theory (Gainesville, FL: Univ. Florida)

Brent, R. P. 1973, Algorithms for Minimization without Derivatives (Englewood Cliffs, NJ: Prentice-Hall)
Coughlin, J. L. 2010, arXiv:1004.1395
Đurašević, G. 1992, Ap&SS, 197, 17
Euler, L. 1755, Institutiones Calculi Differentialis (Berlin: Mémoires de l'académie de sciences)
Galassi, M., Davies, J., Theiler, J., et al. 2003, GNU Scientific Library Reference Manual (2nd ed; Network Theory Ltd.)
Gelman, A., & Rubin, D. B. 1992, StaSc, 7, 457
Hill, G., & Rucinski, S. 1993, in Modeling and Analysis of Eclipsing Binary Observations, Vol. 135, ed. E. F. Milone (New York: Springer)
Hogg, D. W. 2012, arXiv:1205.4446
Hogg, D. W., Bovy, J., & Lang, D. 2010, arXiv:1008.4686
Hogg, D. W., & Foreman-Mackey, D. 2017, arXiv:1710.06068
Ingber, L. 1989, MComM, 12, 967
Ingber, L. 1996, Control and Cybernetics, 25, 33
Irwin, J. B. 1947, ApJ, 106, 380
Kallrath, J., & Linnell, A. P. 1987, ApJ, 313, 346
Levenberg, K. 1944, QApMa, 2, 164
MacKay, D. J. C. 2002, Information Theory, Inference & Learning Algorithms (New York: Cambridge Univ. Press)
Marquardt, D. 1963, SJAM, 11, 431
Metropolis, N., Rosenbluth, A. W., Rosenbluth, M. N., Teller, A. H., & Teller, E. 1953, JChPh, 21, 1087
Nelder, J. A., & Mead, R. 1965, CompJ, 7, 308
Pols, O. R., Tout, C. A., Eggleton, P. P., & Han, Z. 1995, MNRAS, 274, 964
Press, W. H., Teukolsky, S. A., Vetterling, W. T., & Flannery, B. P. 2007, Numerical Recipes: The Art of Scientific Computing (3rd ed.; New York: Cambridge Univ. Press)
Prša, A., & Zwitter, T. 2005, ApJ, 628, 426
Prša, A., & Zwitter, T. 2007, in ASP Conf. Ser. 370, Solar and Stellar Physics through Eclipses, ed. O. Demircan, S. O. Selam, & B. Albayrak (San Francisco, CA: ASP), 175
Rasmussen, C. E., & Williams, C. K. I. 2005, Gaussian Processes for Machine Learning (Adaptive Computation and Machine Learning) (Cambridge, MA: MIT Press)
Serfozo, R. 2009, Basics of Applied Stochastic Processes, Probability and Its Applications (Berlin: Springer)
Torres, G., Andersen, J., & Giménez, A. 2010, A&ARv, 18, 67
Wilson, R. E., & Biermann, P. 1976, A&A, 48, 349
Wilson, R. E., & Devinney, E. J. 1971, ApJ, 166, 605
Wilson, R. E., & Van Hamme, W. 2003, Computing Binary Star Observables (Gainesville, FL: Univ. Florida)
Wyse, A. B., & Kron, G. E. 1939, LicOB, 19, 17

Chapter 9

Conclusion

In this book we covered topics such as timekeeping considerations, basic observational techniques that lend themselves readily to binary stars, and the foundations of geometry and radiation that apply to binaries, as well as several advanced topics, including higher-order multiples and misaligned orbits, and we concluded by presenting a step-by-step procedure for building a model and some considerations for solving the inverse problem. This portfolio of background knowledge enables us to tackle real-world binary modeling. The code that is built on the principles presented in this book, PHOEBE (Prša & Zwitter 2005; Prša et al. 2016), is open source, free, and publicly available from http://phoebe-project.org.

The best starting point to put this knowledge into actual use might be to tackle some of the already-published solutions. The Catalog and AtLas of Eclipsing Binaries (CALEB), maintained by Dave Bradstreet and available from http://caleb.eastern.edu, contains over 300 stars with data and models readily available for download.

The next step is to plunge into large data sets that include binary stars. These include the *Kepler* Eclipsing Binary Catalog (available at http://keplerEBs.villanova.edu); the OGLE IV catalog toward the Galactic bulge (Soszyński et al. 2016); binaries in the All Sky Automated Survey (ASAS), the Northern Sky Variability Survey (NSVS), and the Lincoln Near-Earth Asteroid Research (LINEAR) survey (Lee 2015); and soon *Gaia* and LSST data sets.

To conclude, we give a few brief, subjective pieces of advice. When publishing eclipsing binary results, be mindful of parameter correlations and solution degeneracy. Convince yourself that your solution is unique (within the parameter uncertainties) and that those uncertainties are realistic. Do not assume that the noise in the data is white; instead, add a noise model and assess its validity. Use the modeling code as a tool, not as a black box. Solve binary star data not for the sake of solving them, but to use the solutions to answer specific astrophysical questions. Be wary of automated methods that yield solutions for many binaries at once (this is

doi:10.1088/978-0-7503-1287-5ch9

coming from a person who devised a method based on artificial intelligence); while the *bulk* results are certainly useful, the solutions for individual systems likely suffer from method systematics.

A new era of petabytes of data is upon us. As the available data change, the models need to change as well, especially in light of the ever-growing computer power. This book laid out the foundations of our current understanding of the processes that govern eclipsing binary observables, but many details are yet to be worked out, and I for one feel privileged to be a part of this majestic endeavor.

References

Lee, C.-H. 2015, MNRAS, 453, 3474
Prša, A., & Zwitter, T. 2005, ApJ, 628, 426
Prša, A., Conroy, K. E., Horvat, M., et al. 2016, ApJS, 227, 29
Soszyński, I., Pawlak, M., Pietrukowicz, P., et al. 2016, AcA, 66, 405

Lightning Source UK Ltd.
Milton Keynes UK
UKHW051223260919
350440UK00004B/51/P

9 780750 312882